CONTINUED FRACTIONS

NEW MATHEMATICAL LIBRARY

published by
The Mathematical Association of America

The New Mathematical Library (NML) was begun in 1961 by the School Mathematics Study Group to make available to high school students short expository books on various topics not usually covered in the high school syllabus. In a decade the NML matured into a steadily growing series of some twenty titles of interest not only to the originally intended audience, but to college students and teachers at all levels. Previously published by Random House and L. W. Singer, tne NML became a publication series of the Mathematical Association of America (MAA) in 1975. Under the auspices of the MAA the NML will continue to grow and will remain dedicated to its original and expanded purposes.

CONTINUED FRACTIONS

by

C. D. Olds

San Jose State University

9

MATHEMATICAL ASSOCIATION
OF AMERICA

Illustrations by Carl Bass

Eleventh Printing

 Published in Washington, D.C. by
The Mathematical Association of America

Library of Congress Catalog Card Number: 61-12185

Complete Set ISBN-0-88385-600-X

Volume 22: ISBN-0-88385-609-3

Manufactured in the United States of America

Note to the Reader

This book is one of a series written by professional mathematicians in order to make some important mathematical ideas interesting and understandable to a large audience of high school students and laymen. Most of the volumes in the *New Mathematical Library* cover topics not usually included in the high school curriculum; they vary in difficulty, and, even within a single book, some parts require a greater degree of concentration than others. Thus, while the reader needs little technical knowledge to understand most of these books, he will have to make an intellectual effort.

If the reader has so far encountered mathematics only in classroom work, he should keep in mind that a book on mathematics cannot be read quickly. Nor must he expect to understand all parts of the book on first reading. He should feel free to skip complicated parts and return to them later; often an argument will be clarified by a subsequent remark. On the other hand, sections containing thoroughly familiar material may be read very quickly.

The best way to learn mathematics is to *do* mathematics, and each book includes problems, some of which may require considerable thought. The reader is urged to acquire the habit of reading with paper and pencil in hand; in this way mathematics will become increasingly meaningful to him.

The authors and editorial committee are interested in reactions to the books in this series and hope that readers will write to: Anneli Lax, Editor, New Mathematical Library, NEW YORK UNIVERSITY, THE COURANT INSTITUTE OF MATHEMATICAL SCIENCES, 251 Mercer Street, New York, N. Y. 10012.

The Editors

NEW MATHEMATICAL LIBRARY

1 Numbers: Rational and Irrational *by Ivan Niven*
2 What is Calculus About? *by W. W. Sawyer*
3 An Introduction to Inequalities *by E. F. Beckenbach and R. Bellman*
4 Geometric Inequalities *by N. D. Kazarinoff*
5 The Contest Problem Book I Annual High School Mathematics Examinations 1950–1960. Compiled and with solutions *by Charles T. Salkind*
6 The Lore of Large Numbers, *by P. J. Davis*
7 Uses of Infinity *by Leo Zippin*
8 Geometric Transformations I *by I. M. Yaglom, translated by A. Shields*
9 Continued Fractions *by Carl D. Olds*
10 Graphs and Their Uses *by Oystein Ore*
11 } Hungarian Problem Books I and II, Based on the Eötvös
12 } Competitions 1894–1905 and 1906–1928, *translated by E. Rapaport*
13 Episodes from the Early History of Mathematics *by A. Aaboe*
14 Groups and Their Graphs *by I. Grossman and W. Magnus*
15 The Mathematics of Choice or How to Count Without Counting *by Ivan Niven*
16 From Pythagoras to Einstein *by K. O. Friedrichs*
17 The Contest Problem Book II Annual High School Mathematics Examinations 1961–1965. Compiled and with solutions *by Charles T. Salkind*
18 First Concepts of Topology *by W. G. Chinn and N. E. Steenrod*
19 Geometry Revisited *by H. S. M. Coxeter and S. L. Greitzer*
20 Invitation to Number Theory *by Oystein Ore*
21 Geometric Transformations II *by I. M. Yaglom, translated by A. Shields*
22 Elementary Cryptanalysis—A Mathematical Approach *by A. Sinkov*
23 Ingenuity in Mathematics *by Ross Honsberger*
24 Geometric Transformations III *by I. M. Yaglom, translated by A. Shenitzer*
25 The Contest Problem Book III Annual High School Mathematics Examinations 1966–1972. Compiled and with solutions *by C. T. Salkind and J. M. Earl*
26 Mathematical Methods in Science *by George Pólya*
27 International Mathematical Olympiads—1959–1977. Compiled and with solutions *by S. L. Greitzer*
28 The Mathematics of Games and Gambling *by Edward W. Packel*
29 The Contest Problem Book IV Annual High School Mathematics Examinations 1973–1982. Compiled and with solutions *by R. A. Artino, A. M. Gaglione and N. Shell*
30 The Role of Mathematics in Science *by M. M. Schiffer and L. Bowden*
31 International Mathematical Olympiads 1978–1985 and forty supplementary problems. Compiled and with solutions *by Murray S. Klamkin*
32 Riddles of the Sphinx (and Other Mathematical Puzzle Tales) *by Martin Gardner*

Other titles in preparation.

Contents

CONTINUED FRACTIONS

Preface

At first glance nothing seems simpler or less significant than writing a number, for example $\frac{9}{7}$, in the form

$$\frac{9}{7} = 1 + \frac{2}{7} = 1 + \frac{1}{\frac{7}{2}} = 1 + \cfrac{1}{3 + \cfrac{1}{2}} = 1 + \cfrac{1}{3 + \cfrac{1}{1 + \cfrac{1}{1}}}.$$

It turns out, however, that fractions of this form, called "continued fractions", provide much insight into many mathematical problems, particularly into the nature of numbers.

Continued fractions were studied by the great mathematicians of the seventeenth and eighteenth centuries and are a subject of active investigation today.

Nearly all books on the theory of numbers include a chapter on continued fractions, but these accounts are condensed and rather difficult for the beginner. The plan in this book is to present an easy-going discussion of simple continued fractions that can be understood by anyone who has a minimum of mathematical training.

Mathematicians often think of their subject as a creative art rather than as a science, and this attitude is reflected in the pages that follow. Chapter 1 shows how continued fractions might be discovered accidentally, and then, by means of examples, how rational fractions can be expanded into continued fractions. Gradually more general notation is introduced and preliminary theorems are stated and proved. In Chapter 2 these results are applied to the solution of linear Diophantine equations. This chapter should be easy to read; it is, if anything, more detailed than necessary.

Chapter 3 deals with the expansion of irrational numbers into infinite continued fractions, and includes an introductory discussion of the idea of limits. Here one sees how continued fractions can be used to give better and better rational approximations to irrational numbers. These and later results are closely connected with and supplement similar ideas developed in Niven's book, *Numbers: Rational and Irrational.*

The periodic properties of continued fractions are discussed in Chapter 4. The reader will find this chapter more challenging than the others, but the end results are rewarding. The main part of the chapter develops a proof of Lagrange's theorem that the continued fraction expansion of every quadratic irrational is periodic after a certain stage; this fact is then used as the key to the solution of Pell's equation.

Chapter 5 is designed to give the reader a look into the future, and to suggest further study of the subject. Here the famous theorem of Hurwitz is discussed, and other theorems closely related to it are mentioned.

It goes without saying that one should not "read" a mathematics book. It is better to get out pencil and paper and rewrite the book. A student of mathematics should wrestle with every step of a proof; if he does not understand it in the first round, he should plan to return to it later and tackle it once again until it is mastered. In addition he should test his grasp of the subject by working the problems at the end of the sections. These are mostly of an elementary nature, closely related to the text, and should not present any difficulties. Their answers appear at the end of the book.

The first of the two appendices gives a proof that $x^2 - 3y^2 = -1$ has no solution in integers, and Appendix II is a collection of miscellaneous expansions designed to show how the subject has developed; many of these expansions are difficult to obtain. Finally, there is a short list of references. In the text "Crystal [2]", for example, refers to item 2 listed in the references.

I wish to express my thanks to the School Mathematics Study Group for including this book in the *New Mathematical Library* series, and to the Editorial Panel for suggestions which have improved the book. Particular thanks are due to Dr. Anneli Lax, not only for technical advice, so freely given, but also for her critical reading of the text. I am also grateful to my wife who typed the original manuscript, and to Mrs. Ruth Murray, who prepared the final typescript.

C. D. Olds

Los Altos, California, 1961.

CHAPTER ONE

Expansion of Rational Fractions

1.1 Introduction

Imagine that an algebra student attempts to solve the quadratic equation

$$x^2 - 3x - 1 = 0 \tag{1.1}$$

as follows: He first divides through by x and writes the equation in the form

$$x = 3 + \frac{1}{x}\cdot$$

The unknown quantity x is still found on the right-hand side of this equation and hence can be replaced by its equal, namely $3 + 1/x$. This gives

$$x = 3 + \frac{1}{x} = 3 + \cfrac{1}{3 + \cfrac{1}{x}}\cdot$$

Repeating this replacement of x by $3 + 1/x$ several more times he obtains the expression

$$x = 3 + \cfrac{1}{3 + \cfrac{1}{3 + \cfrac{1}{3 + \cfrac{1}{3 + \cfrac{1}{x}}}}}\cdot \tag{1.2}$$

Since x continues to appear on the right-hand side of this "multiple-decked" fraction, he does not seem to be getting any closer to the solution of the equation (1.1).

But let us look more closely at the right side of equation (1.2). We see that it contains a *succession* of fractions,

$$(1.3) \quad 3, \quad 3+\frac{1}{3}, \quad 3+\cfrac{1}{3+\cfrac{1}{3}}, \quad 3+\cfrac{1}{3+\cfrac{1}{3+\cfrac{1}{3}}}, \quad \cdots,$$

obtained by stopping at consecutive stages. These numbers, when converted into fractions and then into decimals, give in turn the numbers

$$3, \qquad \frac{10}{3} = 3.333\cdots, \qquad \frac{33}{10} = 3.3, \qquad \frac{109}{33} = 3.30303\cdots.$$

It then comes as a very pleasant surprise to discover that these numbers (or *convergents* as we shall call them later) give better and better approximations to the positive root of the given quadratic equation (1.1). The quadratic formula shows that this root is actually equal to

$$x = \frac{3+\sqrt{13}}{2} = 3.302775\cdots,$$

which, when rounded to 3.303, is in agreement to three decimal places with the last result above.

These preliminary calculations suggest some interesting questions. First, if we calculate more and more *convergents* (1.3), will we continue to get better and better approximations to $x = \frac{1}{2}(3+\sqrt{13})$? Second, suppose we consider the process used to get (1.2) as being continued indefinitely, so that we have in place of (1.2) the non-terminating expression

$$(1.4) \qquad x = 3+\cfrac{1}{3+\cfrac{1}{3+\ddots}},$$

where the three dots stand for the words "and so on" and indicate that the successive fractions are continued without end. Then will the expression on the right of (1.4) actually be equal to $\frac{1}{2}(3+\sqrt{13})$? This reminds us of an infinite decimal. For example, what is meant when we say that the infinite decimal $0.333\cdots$ is equal to $\frac{1}{3}$? These and many other questions will eventually be discussed and answered.

Multiple-decked fractions like (1.2) and (1.4) are called *continued fractions.* A study of these fractions and their many properties and applications forms one of the most intriguing chapters in mathematics. We must start with simpler things, however. The first of these is the introduction of basic definitions.

1.2 Definitions and Notation

An expression of the form

$$a_1 + \cfrac{b_1}{a_2 + \cfrac{b_2}{a_3 + \cfrac{b_3}{a_4 + \ddots}}} \tag{1.5}$$

is called a *continued fraction*. In general, the numbers $a_1, a_2, a_3, \cdots, b_1, b_2, b_3, \cdots$ may be any real or complex numbers, and the number of terms may be finite or infinite.

In this monograph, however, we shall restrict our discussion to *simple continued fractions.* These have the form

$$a_1 + \cfrac{1}{a_2 + \cfrac{1}{a_3 + \cfrac{1}{a_4 + \ddots}}}, \tag{1.6}$$

where the first term a_1 is usually a positive or negative integer (but could be zero), and where the terms $a_2, a_3, a_4, \cdots$ are positive integers. In fact, until we come to Chapter 3, we shall further restrict the discussion to *finite simple continued fractions.* These have the form

$$a_1 + \cfrac{1}{a_2 + \cfrac{1}{a_3 + \cfrac{1}{a_4 + \ddots + \cfrac{1}{a_{n-1} + \cfrac{1}{a_n}}}}}, \tag{1.7}$$

with only a finite number of *terms* $a_1, a_2, a_3, \cdots, a_n$. Such a fraction is called a *terminating* continued fraction. From now on, unless the contrary is stated, the words *continued fraction* will imply that we are dealing with a *finite simple continued fraction.*

A much more convenient way of writing (1.7) is

$$a_1 + \frac{1}{a_2 +}\,\frac{1}{a_3 +}\,\frac{1}{a_4 +}\cdots\frac{1}{+\,a_n}, \tag{1.8}$$

where the + signs after the first one are lowered to remind us of the "step-down" process in forming a continued fraction. It is also convenient to denote the continued fraction (1.8) by the symbol $[a_1, a_2, \cdots, a_n]$, so that

$$[a_1, a_2, \cdots, a_n] = a_1 + \frac{1}{a_2 +}\,\frac{1}{a_3 +}\cdots\frac{1}{+\,a_n}. \tag{1.9}$$

The terms $a_1, a_2, \cdots, a_n$ are called the *partial quotients* of the continued fraction.

1.3 Expansion of Rational Fractions

A *rational number* is a fraction of the form p/q where p and q are integers with $q \neq 0$. We shall prove in the next section that *every rational fraction, or rational number, can be expressed as a finite simple continued fraction.*

For example, the continued fraction for $\frac{67}{29}$ is

$$\frac{67}{29} = 2 + \cfrac{1}{3 + \cfrac{1}{4 + \cfrac{1}{2}}} = 2 + \frac{1}{3 +}\,\frac{1}{4 +}\,\frac{1}{2},$$

or

$$\frac{67}{29} = [2, 3, 4, 2].$$

How did we get this result? First we divided 67 by 29 to obtain the quotient 2 and the remainder 9, so that

$$\frac{67}{29} = 2 + \frac{9}{29} = 2 + \cfrac{1}{\cfrac{29}{9}}. \tag{1.10}$$

Note that on the right we have replaced $\frac{9}{29}$ by the reciprocal of $\frac{29}{9}$.

Next we divided 29 by 9 to obtain

$$\frac{29}{9} = 3 + \frac{2}{9} = 3 + \frac{1}{\frac{9}{2}}\cdot \tag{1.11}$$

Finally, we divided 9 by 2 to obtain

$$\frac{9}{2} = 4 + \frac{1}{2}, \tag{1.12}$$

at which stage the process terminates. Now substitute (1.12) into (1.11), and then substitute (1.11) into (1.10) to get

$$\frac{67}{29} = 2 + \frac{1}{\frac{29}{9}} = 2 + \frac{1}{3 + \frac{1}{\frac{9}{2}}} = 2 + \frac{1}{3 + \frac{1}{4 + \frac{1}{2}}},$$

or

$$\frac{67}{29} = [2, 3, 4, 2] = [a_1, a_2, a_3, a_4]. \tag{1.13}$$

We should notice that in equation (1.10) the number $2 \cdot 29$ is the largest multiple of 29 that is less than 67, and consequently the remainder (in this case the number 9) is necessarily a number ≥ 0 but definitely < 29.†

Next consider equation (1.11). Here $3 \cdot 9$ is the largest multiple of 9 that is less than 29. The remainder, 2, is necessarily a number ≥ 0 but < 9.

In (1.12) the number $4 \cdot 2$ is the largest multiple of 2 that is less than 9 and the remainder is 1, a number ≥ 0 but < 2.

Finally, we cannot go beyond equation (1.12), for if we write

$$\frac{9}{2} = 4 + \frac{1}{\frac{2}{1}} = 4 + \frac{1}{2},$$

then $2 \cdot 1$ is the largest multiple of 1 that divides 2 and we simply end up with

$$\frac{2}{1} = 2 \cdot 1 + 0 = 2,$$

so the calculation terminates.

† If a number a is less than a number b we write $a < b$. If a is less than or equal to b we write $a \leq b$. Likewise, if a is greater than b, or if a is greater than or equal to b, we write, respectively, $a > b$, $a \geq b$. For a detailed discussion of inequalities, see E. Beckenbach and R. Bellman [1].

The process for finding the continued fraction expansion for $\frac{67}{29}$ can be arranged as follows:

$29\overline{)67}(2 = a_1$	Divide 67 by 29.
$\quad 58$	$2 \cdot 29 = 58$; subtract 58 from 67.
$\quad 9\overline{)29}(3 = a_2$	Divide 29 by 9.
$\qquad 27$	$3 \cdot 9 = 27$; subtract 27 from 29.
$\qquad 2\overline{)9}(4 = a_3$	Divide 9 by 2.
$\qquad\quad 8$	$4 \cdot 2 = 8$; subtract 8 from 9.
$\qquad\quad 1\overline{)2}(2 = a_4$	Divide 2 by 1.
$\qquad\qquad 2$	$2 \cdot 1 = 2$; subtract 2 from 2.
$\qquad\qquad 0$	Process terminates.

Thus

$$\frac{67}{29} = [a_1, a_2, a_3, a_4] = [2, 3, 4, 2].$$

We observe, in this example, that in the successive divisions the remainders 9, 2, 1 are exactly determined non-negative numbers each smaller than the corresponding divisor. Thus the remainder 9 is less than the divisor 29, the remainder 2 is less than the divisor 9, and so on. The remainder in each division becomes the divisor in the next division, so that the successive remainders become smaller and smaller non-negative integers. Thus the remainder zero must be reached eventually, and the process must end.

Each remainder obtained in this process is a *unique* non-negative number. For example, can you divide 67 by 29, obtain the largest quotient 2, and end up with a remainder other than 9? This means that, for the given fraction $\frac{67}{29}$, our process yields exactly one sequence of remainders.

As a second example, let us find the continued fraction expansion for $\frac{29}{67}$. We obtain

$$67\overline{)29}(0 = a_1$$
$$0$$
$$29\overline{)67}(2 = a_2$$
$$58$$
$$9\overline{)29}(3 = a_3$$
$$27$$
$$2\overline{)9}(4 = a_4$$
$$8$$
$$1\overline{)2}(2 = a_5.$$
$$2$$
$$0$$

Hence

$$\frac{29}{67} = [0, 2, 3, 4, 2] = [a_1, a_2, a_3, a_4, a_5].$$

Notice that in this example $a_1 = 0$. To check our results, all we have to do is simplify the continued fraction

$$0 + \cfrac{1}{2 + \cfrac{1}{3 + \cfrac{1}{4 + \cfrac{1}{2}}}} = \cfrac{1}{2 + \cfrac{1}{3 + \cfrac{2}{9}}} = \cfrac{1}{2 + \cfrac{9}{29}} = \frac{29}{67}.$$

A comparison of the expansion $\frac{67}{29} = [2, 3, 4, 2]$ with the expansion of the reciprocal $\frac{29}{67} = [0, 2, 3, 4, 2]$ suggests the result that, if p is greater than q and

$$\frac{p}{q} = [a_1, a_2, \cdots, a_n],$$

then

$$\frac{q}{p} = [0, a_1, a_2, \cdots, a_n].$$

The reader is asked to state a similar result for $p < q$.

The following examples will help to answer some questions which may have occurred to the attentive student.

First, is the expansion

$$\frac{67}{29} = [2, 3, 4, 2] = [a_1, a_2, a_3, a_4]$$

the *only* expansion of $\frac{67}{29}$ as a simple finite continued fraction? If we go back and study the method by which the expansion was obtained, the answer would seem to be "yes". And this would be true except that a slight change can always be made in the *last* term, or last partial quotient, a_4. Since $a_4 = 2$, we can write

$$\frac{1}{a_4} = \frac{1}{2} = \cfrac{1}{1 + \cfrac{1}{1}}.$$

Hence it is also true that

$$\frac{67}{29} = [2, 3, 4, 1, 1].$$

Clearly the expansion [2, 3, 4, 1, 1] can be changed back to its original form [2, 3, 4, 2]. We shall see in the more general discussion which follows that this is the only way we can get a "different" expansion.

Next, let us consider how to obtain the expansion of a negative rational number $-p/q$. This requires only a slight variation of the process already explained. For example, to find the continued fraction expansion of $-\frac{37}{44}$, proceed as follows:

```
44)-37(-1
   -44
    7)44(6
      42
       2)7(3
         6
         1)2(2 .
           2
           0
```

(Search for a negative quotient which, when multiplied by 44 and subtracted from -37, leaves the *smallest positive* remainder.)

Thus

$$-\frac{37}{44} = [-1, 6, 3, 2] = [-1, 6, 3, 1, 1] = [a_1, a_2, a_3, a_4, a_5].$$

Notice that a_1 is negative, but a_2, a_3, a_4, a_5 are positive.

The third question is this: If we multiply the numerator and denominator of $\frac{67}{29}$ by some number, say 3, and then expand the resulting fraction, $\frac{201}{87}$, will the continued fraction for $\frac{201}{87}$ be the same as that for $\frac{67}{29}$? We shall see that the expansions are identical, for

(1.14)

```
87)201(2
   174
    27)87(3
       81
        6)27(4
          24
           3)6(2 .
             6
             0
```

Thus

$$\frac{201}{87} = \frac{67}{29} = [2, 3, 4, 2].$$

This illustrates an interesting property of continued fractions. If we calculated

$$[2, 3, 4, 2] = 2 + \frac{1}{3} + \frac{1}{4} + \frac{1}{2},$$

we would get back to $\frac{67}{29}$, *not* to $\frac{201}{87}$. We always obtain a rational fraction p/q in its *lowest terms,* i.e., a fraction for which p and q *have no factors greater than* 1 *in common.* Can you discover at this stage a reason for this? Later an explanation will be given.

Problem Set 1

1. Convert each of the following into finite simple continued fractions.

(a) $\frac{17}{11}$ (e) $.23 = \frac{23}{100}$

(b) $\frac{51}{33}$ (f) $\frac{355}{106}$

(c) $3.54 = \frac{354}{100}$ (g) 3.14159

(d) $\frac{233}{177}$

2. Find p/q if

$$\frac{p}{q} = 3 + \frac{1}{4} + \frac{1}{1} + \frac{1}{5}.$$

3. Find p/q if $p/q = [0, 2, 1, 4, 2]$.

4. Find p/q if $p/q = [3, 7, 15, 1]$. Convert p/q to a decimal and compare with the value of π.

5. Find the simple continued fraction expansions of (a) $\frac{11}{17}$, (b) $\frac{33}{51}$; compare these with the expansions in Problem 1 (a), (b).

6. Show that, if $p > q$ and $p/q = [a_1, a_2, \cdots, a_n]$, then $q/p = [0, a_1, a_2, \cdots, a_n]$; and conversely, if $q/p = [0, a_1, a_2, \cdots, a_n]$ then $p/q = [a_1, a_2, \cdots, a_n]$.

1.4 Expansion of Rational Fractions (General Discussion)

So far we have introduced the terminology peculiar to the study of continued fractions and have worked with particular examples. But to make real progress in our study we must discuss more general results. Working with symbols instead of with actual numbers frees the mind and allows us to think abstractly. Thus, while our first theorem merely expresses in general terms what we did in the worked examples, once this has been accomplished a host of other ideas quickly follows.

THEOREM 1.1. *Any finite simple continued fraction represents a rational number. Conversely, any rational number p/q can be represented as a finite simple continued fraction; with the exceptions to be noted below, the representation, or expansion, is unique.*

PROOF. The first sentence in this theorem is quite clear from what we have explained in our worked examples, for if any expansion terminates we can always "back track" and build the expansion into a rational fraction.

To prove the converse, let p/q, $q > 0$, be any rational fraction. We divide p by q to obtain

$$\frac{p}{q} = a_1 + \frac{r_1}{q}, \qquad 0 \le r_1 < q,$$

where a_1 is the unique integer so chosen as to make the remainder r_1 greater than or equal to 0 and less than q. As we saw in the worked examples, a_1 can be negative, zero, or positive. If $r_1 = 0$, the process terminates and the continued fraction expansion for p/q is $[a_1]$.

If $r_1 \neq 0$, we write

$$\frac{p}{q} = a_1 + \cfrac{1}{\cfrac{q}{r_1}}, \qquad 0 < r_1 < q, \tag{1.15}$$

and repeat the division process, dividing q by r_1 to obtain

$$\frac{q}{r_1} = a_2 + \frac{r_2}{r_1}, \qquad 0 \le r_2 < r_1. \tag{1.16}$$

Notice now that q/r_1 is a positive fraction, so a_2 is the unique largest positive integer that makes the remainder r_2 a number between 0 and r_1. If $r_2 = 0$, the process stops and we substitute $q/r_1 = a_2$ from (1.16) into (1.15) to obtain

$$\frac{p}{q} = a_1 + \frac{1}{a_2} = [a_1, a_2]$$

as the continued fraction expansion for p/q.

If $r_2 \neq 0$, we write (1.16) in the form

$$\frac{q}{r_1} = a_2 + \cfrac{1}{\cfrac{r_1}{r_2}}, \qquad 0 < r_2 < r_1, \tag{1.17}$$

and repeat the division process using r_1/r_2.

We observe that the calculations stop when we come to a remainder $r_n = 0$. Is it possible never to arrive at an r_n which is zero, so that the division process continues indefinitely? This is clearly impossible, for the remainders $r_1, r_2, r_3, \cdots$ form a decreasing sequence of non-negative integers $q > r_1 > r_2 > r_3 > \cdots$ and unless we come eventually to a remainder r_n which is equal to zero, we shall be in the ridiculous position of having discovered an infinite number of distinct positive integers all less than a finite positive integer q.

Hence, by successive divisions we obtain a sequence of equations:

$$\begin{aligned}
\frac{p}{q} &= a_1 + \frac{r_1}{q}, & 0 &< r_1 < q,\\
\frac{q}{r_1} &= a_2 + \frac{r_2}{r_1}, & 0 &< r_2 < r_1,\\
\frac{r_1}{r_2} &= a_3 + \frac{r_3}{r_2}, & 0 &< r_3 < r_2,\\
&\cdots\cdots\cdots & &\cdots\cdots\\
\frac{r_{n-3}}{r_{n-2}} &= a_{n-1} + \frac{r_{n-1}}{r_{n-2}}, & 0 &< r_{n-1} < r_{n-2},\\
\frac{r_{n-2}}{r_{n-1}} &= a_n + \frac{0}{r_{n-1}} = a_n + 0, & r_n &= 0,
\end{aligned} \tag{1.18}$$

terminating, after a certain finite number of divisions, with the equation in which the remainder r_n is equal to zero.

It is now easy to represent p/q as a finite simple continued fraction. From the first two equations in (1.18) we have

$$\frac{p}{q} = a_1 + \cfrac{1}{\cfrac{q}{r_1}} = a_1 + \cfrac{1}{a_2 + \cfrac{1}{\cfrac{r_1}{r_2}}}.$$

Using the third equation in (1.18) we replace r_1/r_2 by

$$a_3 + \cfrac{1}{\cfrac{r_2}{r_3}},$$

and so on, until finally we obtain the expansion

$$(1.19) \quad \frac{p}{q} = a_1 + \frac{1}{a_2 +}\,\frac{1}{a_3 +}\cdots\frac{1}{+\,a_n} = [a_1, a_2, a_3, \cdots, a_n].$$

The *uniqueness* of the expansion (1.19) follows from the manner in which the a_i's are calculated. This statement must be accompanied, however, by the remark that once the expansion has been obtained we can always modify the *last* term a_n so that the number of terms in the expansion is either *even* or *odd*, as we choose. To see this, notice that if a_n is greater than 1 we can write

$$\frac{1}{a_n} = \cfrac{1}{(a_n - 1) + \cfrac{1}{1}},$$

so that (1.19) can be replaced by

$$(1.20) \qquad \frac{p}{q} = [a_1, a_2, \cdots, a_{n-1}, a_n - 1, 1].$$

On the other hand, if $a_n = 1$, then

$$\cfrac{1}{a_{n-1} + \cfrac{1}{a_n}} = \frac{1}{(a_{n-1} + 1)},$$

so that (1.19) becomes

$$(1.21) \qquad \frac{p}{q} = [a_1, a_2, \cdots, a_{n-2}, a_{n-1} + 1].$$

Hence we have the following theorem:

THEOREM 1.2. *Any rational number p/q can be expressed as a finite simple continued fraction in which the last term can be modified so as to make the number of terms in the expansion either even or odd.*

It is interesting to notice that the equations (1.18) are precisely the equations used in a procedure known as *Euclid's algorithm* for finding the *greatest common divisor* of the integers p and q. † [This procedure occurs in the seventh book of Euclid's *Elements* (about 300 B.C.); however it is known to be of earlier origin.]

To find the greatest common divisor of p and q by means of *Euclid's algorithm*, we write the equations (1.18) in the form:

$$\begin{array}{llr} & p = a_1 q + r_1, & 0 < r_1 < q, \\ & q = a_2 r_1 + r_2, & 0 < r_2 < r_1, \\ & r_1 = a_3 r_2 + r_3, & 0 < r_3 < r_2, \\ (1.22) & \dots\dots\dots\dots & \dots\dots\dots \\ & r_{n-3} = a_{n-1} r_{n-2} + r_{n-1}, & 0 < r_{n-1} < r_{n-2}, \\ & r_{n-2} = a_n r_{n-1} + 0 = a_n r_{n-1}, & 0 = r_n. \end{array}$$

The first equation, $p = a_1 q + r_1$, is obtained from the first equation in (1.18) by multiplying both sides by the denominator q; similarly for the other equations.

We shall prove that the last *nonvanishing* remainder r_{n-1} is the g.c.d. of p and q. In order to do this, we first state the two conditions that the g.c.d. of two integers must satisfy. The number d is the g.c.d. of two integers p and q if

(a) d divides both integers p and q, and
(b) any common divisor c of p and q divides d.

For example, let $p = 3 \cdot 5 \cdot 11$ and let $q = 3^2 \cdot 5 \cdot 13$. Then the g.c.d. of p and q is $d = 3 \cdot 5$, since (a) $d = 3 \cdot 5$ divides both p and q; and (b) the common divisors 3 and 5 of p and q divide d.

We need only one more observation: If a, b, and c are integers such that

$$a = b + c,$$

then any integer d which divides both a and b must divide c. For if d divides a, then $a = da_1$ where a_1 is an integer, and if d divides b, then $b = db_1$, b_1 an integer. Since $a - b = c$, we see that

$$a - b = da_1 - db_1 = d(a_1 - b_1) = c,$$

so that d divides c. Likewise, any integer d which divides both b and c will also divide a.

† The *greatest common divisor* (g.c.d.) of any two integers p and q is the largest integer which divides both p and q. In the theory of numbers the g.c.d. of the integers p and q is denoted by the symbol (p, q); thus, $(p, q) = d$ means that d is the largest integral factor common to both p and q.

We now return to the equations (1.22). The last equation there,

$$r_{n-2} = a_n r_{n-1},$$

shows that r_{n-1} divides, or is a factor of, r_{n-2}. The equation directly above it, namely

$$r_{n-3} = a_{n-1} r_{n-2} + r_{n-1},$$

shows that r_{n-1} divides r_{n-3}, since it divides r_{n-1} and r_{n-2}. In the same way, from the equation

$$r_{n-4} = a_{n-2} r_{n-3} + r_{n-2},$$

we see that r_{n-1} divides r_{n-4}, since it divides both r_{n-2} and r_{n-3}. Working up from the bottom in this fashion, we find that r_{n-1} divides r_3 and r_2, and hence divides r_1. Dividing r_2 and r_1, it divides q; and finally, dividing both r_1 and q, it divides p. Hence, r_{n-1} divides both p and q, and condition (a) is satisfied.

Next we must show that if c is any common divisor of both p and q, then c divides r_{n-1}. This time we start with the first equation in (1.22) and work our way down. If c divides both p and q, the first equation in (1.22) shows that c divides r_1. But if c divides both q and r_1, the second equation in (1.22) shows that c divides r_2. Continuing in this manner, we arrive at the next to the last equation,

$$r_{n-3} = a_{n-1} r_{n-2} + r_{n-1},$$

in which c divides r_{n-3} and r_{n-2}, and hence divides r_{n-1}. Thus condition (b) is satisfied, and we conclude that r_{n-1} is the g.c.d. of p and q.

As an example, let us use Euclid's algorithm to determine the g.c.d. of $p = 6381$ and $q = 5163$. We find that

$$\begin{aligned}
6381 &= 1 \cdot 5163 + 1218\\
5163 &= 4 \cdot 1218 + 291\\
1218 &= 4 \cdot 291 + 54\\
291 &= 5 \cdot 54 + 21\\
54 &= 2 \cdot 21 + 12\\
21 &= 1 \cdot 12 + 9\\
12 &= 1 \cdot 9 + 3\\
9 &= 3 \cdot 3 + 0;
\end{aligned}$$

hence 3 is the g.c.d. of 6381 and 5163. Actually, $6381 = 3^2 \cdot 709$, where 709 is a prime number, and $5163 = 3 \cdot 1721$ where 1721 is also a prime number. (A *prime number* is a number with precisely two positive integral divisors: 1 and the number itself.) Thus 3 is the only factor common to these two numbers, and hence is the g.c.d.

Problem Set 2

1. Expand the following rational fractions into finite simple continued fractions with an even number of terms and also with an odd number of terms:

   (a) $\frac{29}{5}$ (c) $-\frac{29}{5}$ (e) $-\frac{123}{31}$

   (b) $\frac{5}{29}$ (d) $\frac{123}{31}$ (f) $\frac{31}{123}$

2. Use Euclid's algorithm to find the greatest common divisor (g.c.d.) of the following pairs of numbers:

   (a) 1380, 1449 (b) 1517, 2015 (c) 2299, 3800 (d) 3528, 7455

1.5 Convergents and Their Properties

Continued fractions are of great service in solving many interesting problems, but before we can put them to effective use we must study some of their properties in greater detail.

In Section 1.4 we saw that any rational fraction p/q could be expanded into a finite simple continued fraction

$$\frac{p}{q} = [a_1, a_2, \cdots, a_{n-1}, a_n], \tag{1.23}$$

where a_1 is a positive or negative integer, or zero, and where $a_2, a_3, \cdots, a_n$ are positive integers. From now on we will call the numbers $a_1, a_2, \cdots, a_n$ the *partial quotients* or *quotients* of the continued fraction. From these we can form the fractions

$$c_1 = \frac{a_1}{1}, \qquad c_2 = a_1 + \frac{1}{a_2}, \qquad c_3 = a_1 + \frac{1}{a_2 + \frac{1}{a_3}}, \qquad \cdots,$$

obtained, in succession, by cutting off the expansion process after the first, second, third, $\cdots$ steps. These fractions are called the first, second, third, $\cdots$ *convergents*, respectively, of the continued fraction (1.23). The nth convergent,

$$c_n = a_1 + \frac{1}{a_2 +} \cdots \frac{1}{+ a_n} = [a_1, a_2, \cdots, a_n],$$

it equal to the continued fraction itself.

It is important to develop a systematic way of computing these convergents. We write

$$c_1 = \frac{a_1}{1} = \frac{p_1}{q_1},$$

where $p_1 = a_1$, $q_1 = 1$. Next we write

$$c_2 = a_1 + \frac{1}{a_2} = \frac{a_1a_2 + 1}{a_2} = \frac{p_2}{q_2},$$

where $p_2 = a_1a_2 + 1$ and $q_2 = a_2$; then

$$c_3 = a_1 + \frac{1}{a_2 +} \frac{1}{a_3} = \frac{a_1a_2a_3 + a_1 + a_3}{a_2a_3 + 1} = \frac{p_3}{q_3},$$

$$c_4 = a_1 + \frac{1}{a_2 +} \frac{1}{a_3 +} \frac{1}{a_4} = \frac{a_1a_2a_3a_4 + a_1a_2 + a_1a_4 + a_3a_4 + 1}{a_2a_3a_4 + a_2 + a_4} = \frac{p_4}{q_4},$$

and so on.

Now let us take a closer look at the convergent c_3. We notice that

$$c_3 = \frac{a_3(a_1a_2 + 1) + a_1}{a_3(a_2) + 1} = \frac{a_3p_2 + p_1}{a_3q_2 + q_1} = \frac{p_3}{q_3},$$

so that

$$\text{(1.24)} \quad \begin{aligned} p_3 &= a_3p_2 + p_1 \quad (= a_3a_1a_2 + a_3 + a_1), \\ q_3 &= a_3q_2 + q_1 \quad (= a_3a_2 + 1). \end{aligned}$$

Again, from c_4 we observe, by factoring, that

$$c_4 = \frac{a_4(a_1a_2a_3 + a_1 + a_3) + (a_1a_2 + 1)}{a_4(a_2a_3 + 1) + (a_2)} = \frac{a_4p_3 + p_2}{a_4q_3 + q_2} = \frac{p_4}{q_4},$$

so that

$$\text{(1.25)} \quad \begin{aligned} p_4 &= a_4p_3 + p_2, \\ q_4 &= a_4q_3 + q_2. \end{aligned}$$

From (1.24) and (1.25) we might guess that if

$$c_5 = [a_1, a_2, \cdots, a_5] = \frac{p_5}{q_5},$$

then

$$(1.26) \qquad \begin{aligned} p_5 &= a_5 p_4 + p_3, \\ q_5 &= a_5 q_4 + q_3, \end{aligned}$$

and that in general, for $i = 3, 4, 5, \cdots, n$,

$$c_i = [a_1, a_2, \cdots, a_i] = \frac{p_i}{q_i},$$

where

$$(1.27) \qquad \begin{aligned} p_i &= a_i p_{i-1} + p_{i-2}, \\ q_i &= a_i q_{i-1} + q_{i-2}. \end{aligned}$$

That the equations (1.26) are correct can be confirmed by a direct calculation. This, of course, would not give us proof that the equations (1.27) are true for $i = 3, 4, 5, \cdots, n$, but it is a genuine example of inductive thinking. We guess the formulas from the first few calculations; then, although convinced of their correctness, we must still supply a formal proof. Thus we state and then prove by induction the following theorem:

THEOREM 1.3. *The numerators* p_i *and the denominators* q_i *of the* ith *convergent* c_i *of the continued fraction* $[a_1, a_2, \ldots . a_n]$ *satisfy the equations*

$$(1.28) \qquad \begin{aligned} p_i &= a_i p_{i-1} + p_{i-2}, \\ q_i &= a_i q_{i-1} + q_{i-2}, \end{aligned} \qquad (i = 3, 4, 5, \cdots, n)$$

with the initial values

$$(1.29) \qquad \begin{aligned} p_1 &= a_1, \quad p_2 = a_2 a_1 + 1, \\ q_1 &= 1, \quad q_2 = a_2. \end{aligned}$$

PROOF. We have seen already that $c_1 = p_1/q_1 = a_1/1$ and that $c_2 = p_2/q_2 = (a_2 a_1 + 1)/a_2$. If we substitute $i = 3$ in equations (1.28) we get

$$c_3 = \frac{p_3}{q_3} = \frac{a_3 p_2 + p_1}{a_3 q_2 + q_1} = \frac{a_3(a_2 a_1 + 1) + a_1}{a_3(a_2) + 1},$$

again in agreement with the direct calculation of c_3. Let us assume that Theorem 1.3 is true, or has been verified by direct calculation, for the integers $3, 4, 5, \cdots$ up to some integer k; that is, that

$$(1.30) \qquad c_j = [a_1, a_2, \cdots, a_{j-1}, a_j] = \frac{p_j}{q_j} = \frac{a_j p_{j-1} + p_{j-2}}{a_j q_{j-1} + q_{j-2}},$$

for $j = 3, 4, 5, \cdots, k-1, k$. On the basis of this assumption, we wish to prove that Theorem 1.3 necessarily holds for the next integer $k+1$. To do this we use equations (1.30) to help us supply a proof that

$$(1.31) \quad c_{k+1} = [a_1, a_2, \cdots, a_k, a_{k+1}] = \frac{a_{k+1}p_k + p_{k-1}}{a_{k+1}q_k + q_{k-1}} = \frac{p_{k+1}}{q_{k+1}}.$$

The next few steps will require concentration. Notice first that c_{k+1} differs from c_k only in having $(a_k + 1/a_{k+1})$ in place of a_k. To see this, simply compare

$$c_k = a_1 + \frac{1}{a_2 +}\,\frac{1}{a_3 +}\cdots\frac{1}{+\,a_{k-1} +}\,\frac{1}{a_k}$$

with

$$c_{k+1} = a_1 + \frac{1}{a_2 +}\,\frac{1}{a_3 +}\cdots\frac{1}{+\,a_{k-1} +}\,\frac{1}{\left(a_k + \dfrac{1}{a_{k+1}}\right)}.$$

This suggests that we should be able to calculate c_{k+1} from the formula for c_k obtained from (1.30) with j replaced by k, that is, from

$$(1.32) \qquad c_k = [a_1, a_2, \cdots, a_{k-1}, a_k] = \frac{p_k}{q_k} = \frac{a_k p_{k-1} + p_{k-2}}{a_k q_{k-1} + q_{k-2}}.$$

This we could certainly do if we were sure that the numbers $p_{k-2}, q_{k-2}, p_{k-1}, q_{k-1}$ did not change their values when we tamper with a_k.

To see that they do not, let us look at the manner in which they are calculated. In equation (1.30), first replace j by $k-2$, and then by $k-1$. We obtain in succession:

$$\frac{p_{k-2}}{q_{k-2}} = \frac{a_{k-2}p_{k-3} + p_{k-4}}{a_{k-2}q_{k-3} + q_{k-4}},$$

and

$$\frac{p_{k-1}}{q_{k-1}} = \frac{a_{k-1}p_{k-2} + p_{k-3}}{a_{k-1}q_{k-2} + q_{k-3}}.$$

We notice that the numbers p_{k-1}, q_{k-1} depend only upon the number a_{k-1} and the numbers p_{k-2}, q_{k-2}, p_{k-3}, q_{k-3}, all of which in turn depend upon *preceding* a's, p's, and q's. Thus the numbers p_{k-2}, q_{k-2}, p_{k-1}, q_{k-1} depend only upon the first $k-1$ quotients $a_1, a_2, \cdots, a_{k-1}$ and hence are *independent* of a_k. This means that they will not change when a_k is replaced by $(a_k + 1/a_{k+1})$.

We are now ready to calculate c_{k+1}. In (1.32) replace a_k by $(a_k + 1/a_{k+1})$ to obtain, as we have explained,

$$c_{k+1} = \left[a_1, a_2, \cdots, a_{k-1}, \left(a_k + \frac{1}{a_{k+1}}\right)\right]$$

$$= \frac{\left(a_k + \dfrac{1}{a_{k+1}}\right) p_{k-1} + p_{k-2}}{\left(a_k + \dfrac{1}{a_{k+1}}\right) q_{k-1} + q_{k-2}}.$$

Now, multiplying numerator and denominator by a_{k+1}, we obtain

$$c_{k+1} = \frac{(a_k a_{k+1} + 1)p_{k-1} + a_{k+1}p_{k-2}}{(a_k a_{k+1} + 1)q_{k-1} + a_{k+1}q_{k-2}},$$

and rearranging the terms, we get

$$c_{k+1} = \frac{a_{k+1}(a_k p_{k-1} + p_{k-2}) + p_{k-1}}{a_{k+1}(a_k q_{k-1} + q_{k-2}) + q_{k-1}}.$$

At this point we use the assumption that formulas (1.30) hold for $j = k$, i.e., that

$$a_k p_{k-1} + p_{k-2} = p_k,$$

$$a_k q_{k-1} + q_{k-2} = q_k.$$

Hence the terms in parentheses in the numerator and denominator of our last expression for c_{k+1} can be replaced, respectively, by p_k and q_k. Thus, we obtain

$$c_{k+1} = \frac{a_{k+1}p_k + p_{k-1}}{a_{k+1}q_k + q_{k-1}} = \frac{p_{k+1}}{q_{k+1}}.$$

We have proved, then, that if the expression for the convergent c_j, given by (1.30), holds for the values $j = 3, 4, 5, \cdots, k$, then it also holds for the next convergent $c_{k+1} = p_{k+1}/q_{k+1}$. But we actually know by a direct calculation that (1.30) holds for $j = k = 3$. Hence it is true for the next integer $k + 1 = 4$, and likewise for $k = 5, 6, 7, \cdots, n$. This proves Theorem 1.3.

In studying this proof, notice that nowhere have we used the fact that the quotients a_i are integers. Although each a_i is an integer, the number $a_k + 1/a_k$ need not be one. Nevertheless its substitution for a_k in the proof causes no breakdown of the argument.

It would be convenient if the equations (1.28) could also reproduce the first two convergents given by (1.29). If we put $i = 1, 2$ in (1.28) we get the *undefined* terms p_0, p_{-1}, q_0, q_{-1}. However, if we *assign* the values

$$
\begin{aligned}
p_0 &= 1, \qquad & p_{-1} &= 0, \\
q_0 &= 0, \qquad & q_{-1} &= 1
\end{aligned} \tag{1.33}
$$

to these undefined terms, then equations (1.28) will hold for $i = 1, 2, 3, \cdots, n-1, n$, and the first two values, $i = 1, 2$, will reproduce equations (1.29). Setting i equal to 1 in (1.28), and using (1.33), we get

$$c_1 = \frac{p_1}{q_1} = \frac{a_1 p_0 + p_{-1}}{a_1 q_0 + q_{-1}} = \frac{a_1 1 + 0}{a_1 0 + 1} = \frac{a_1}{1};$$

for $i = 2$ we get

$$c_2 = \frac{p_2}{q_2} = \frac{a_2 p_1 + p_0}{a_2 q_1 + q_0} = \frac{a_2 a_1 + 1}{a_2 1 + 0} = \frac{a_2 a_1 + 1}{a_2}.$$

Hence, the assigned values (1.33) enable us to dispense with equations (1.29) and to use instead equations (1.28), with $i = 1, 2, \cdots, n$. But notice that p_{-1}/q_{-1} and p_0/q_0 are not convergents.

The calculation of successive convergents can now be systematized. An example will make this clear. The continued fraction expansion for $\frac{120}{49}$ is

$$\frac{120}{49} = [2, 2, 4, 2, 2] = [a_1, a_2, a_3, a_4, a_5].$$

We form the following table:

TABLE 1

i	-1	0	1	2	3	4	5
a_i			2	2	4	2	2
p_i	0 ←	1	← 2 ←	5	22	49	120
q_i	1	0	1	2 ←	9	20	49
$c_i = \frac{p_i}{q_i}$			$\frac{2}{1}$	$\frac{5}{2}$	$\frac{22}{9}$	$\frac{49}{20}$	$\frac{120}{49}$

$p_i = a_i p_{i-1} + p_{i-2}, \qquad q_i = a_i q_{i-1} + q_{i-2}, \qquad i = 1, 2, \cdots$

Explanation of table: The entries in the first row of the table are the values of i: $i = -1, 0, 1, 2, \cdots$. Under each value of i the corresponding values of a_i, p_i, q_i, c_i have been listed. Thus, under $i = 4$ we find $a_4 = 2$, $p_4 = 49$, $q_4 = 20$, $c_4 = \frac{49}{20}$.

We form our table in this way: We write the values a_i in the second row, under the values of i to which they correspond. The special values $p_{-1} = 0$, $p_0 = 1$, $q_{-1} = 1$, $q_0 = 0$ are entered at the left, under $i = -1$, $i = 0$, respectively. Then we calculate the p_i's. First, from equations (1.28), using $i = 1$, we get

$$p_1 = a_1p_0 + p_{-1} = 2 \cdot 1 + 0 = 2.$$

(Follow the first system of arrows $\begin{array}{cc} & 2 \\ & \swarrow \\ 0 \leftarrow 1 & \end{array}$.)

We record $p_1 = 2$ under $i = 1$ in the third row. For $i = 2$, we obtain

$$p_2 = a_2p_1 + p_0 = 2 \cdot 2 + 1 = 5, \qquad \begin{array}{cc} & 2 \\ & \swarrow \\ 1 \leftarrow 2 & \end{array},$$

which is recorded under $i = 2$ in the same row. For $i = 3$,

$$p_3 = a_3p_2 + p_1 = 4 \cdot 5 + 2 = 22, \qquad \begin{array}{cc} & 4 \\ & \swarrow \\ 2 \leftarrow 5 & \end{array},$$

and so on. To calculate the q_i's we follow the same scheme, entering the values we obtain in the row labeled q_i. Thus, for example,

$$q_4 = a_4q_3 + q_2 = 2 \cdot 9 + 2 = 20, \qquad \begin{array}{cc} & 2 \\ & \swarrow \\ 2 \leftarrow 9 & \end{array},$$

so 20 is recorded in the fourth row, under $i = 4$.

Problem Set 3

Note: Starred problems are more difficult and could be omitted the first time over.

1. Expand the following rational numbers into simple continued fractions and calculate the successive convergents c_i for each number.

   (a) $\frac{121}{21}$ (b) $\frac{290}{81}$ (c) $\frac{177}{292}$ (d) $\frac{126}{23}$

2. Express each of the following continued fractions in an equivalent form but with an odd number of partial quotients.

   (a) [2, 1, 1, 4, 1, 1] (c) [0, 4, 2, 6]
   (b) [4, 2, 1, 7, 7, 1] (d) [4, 2, 6, 1]

3. For each continued fraction in Problem 2, let n be the number of partial quotients and calculate $p_n q_{n-1} - p_{n-1} q_n$; then calculate the corresponding quantity after these fractions have been expressed with an odd number of partial quotients. In 2 (a), for example, take $p_n/q_n = p_6/q_6$, the last convergent.

4. Calculate the convergents of the continued fraction $[1, 2, 3, 4, 5, 6]$ and show that $p_6 = 5p_5 + 5p_4 + 4p_3 + 3p_2 + 2p_1 + 2$. (See Problem 8 below.)

5. For $[3, 1, 4, 1, 5]$, calculate p_5 and p_4. Then convert p_5/p_4 into a simple continued fraction and compare it with the original fraction. Do the same with q_5/q_4. (See Problem 7.)

6. Calculate the successive convergents to the following approximations to the numbers in parentheses.

   (a) 3.14159 (π) (c) 0.4771 ($\log_{10} 3$)
   (b) 2.718 (e) (d) 0.3010 ($\log_{10} 2$)

*7. Prove that, if $a_1 \neq 0$, then

$$\frac{p_n}{p_{n-1}} = [a_n, a_{n-1}, a_{n-2}, \cdots, a_1],$$

and

$$\frac{q_n}{q_{n-1}} = [a_n, a_{n-1}, a_{n-2}, \cdots, a_2].$$

Hints: We know that $p_n = a_n p_{n-1} + p_{n-2}$; hence

$$\frac{p_n}{p_{n-1}} = a_n + \cfrac{1}{\cfrac{p_{n-1}}{p_{n-2}}}.$$

We also know that $p_{n-1} = a_{n-1} p_{n-2} + p_{n-3}$; hence

$$\frac{p_{n-1}}{p_{n-2}} = a_{n-1} + \cfrac{1}{\cfrac{p_{n-2}}{p_{n-3}}},$$

and so on.

*8. Generalize Problem 4. If $p_1/q_1, p_2/q_2, \cdots, p_n/q_n$ are the convergents of $[1, 2, 3, 4, \cdots, n]$, show that

$$p_n = (n-1)p_{n-1} + (n-1)p_{n-2} + (n-2)p_{n-3} + \cdots + 3p_2 + 2p_1 + (p_1 + 1).$$

Hint: In the relation $p_i = ip_{i-1} + p_{i-2}$, let i be equal to $1, 2, 3, \cdots, n$ and add the resulting expressions. Note that $a_n = n$.

1.6 Differences of Convergents

Those who worked the preceding exercises will already have guessed that the convergents to a finite simple continued fraction are always in their lowest terms. This is a corollary to the following fundamental theorem.

THEOREM 1.4. *If* $p_i = a_ip_{i-1} + p_{i-2}$ *and* $q_i = a_iq_{i-1} + q_{i-2}$ *are defined as in Theorem* 1.3, *then*

$$p_iq_{i-1} - p_{i-1}q_i = (-1)^i, \qquad \textit{where } i \geq 0.$$

PROOF: Direct calculations show that the theorem is true for $i = 0, 1, 2$. When $i = 0$,

$$p_0q_{-1} - p_{-1}q_0 = 1 \cdot 1 - 0 \cdot 0 = 1 = (-1)^0;$$

when $i = 1$,

$$p_1q_0 - p_0q_1 = a_1 \cdot 0 - 1 \cdot 1 = (-1)^1;$$

when $i = 2$,

$$p_2q_1 - p_1q_2 = (a_2a_1 + 1) \cdot 1 - a_1a_2 = 1 = (-1)^2.$$

We shall prove that if the theorem holds for $i = k$, then it also holds for the next integer, $i = k + 1$. From Theorem 1.3 [see equations (1.28)] we know that for $i = k + 1$,

$$p_{k+1} = a_{k+1}p_k + p_{k-1}, \qquad q_{k+1} = a_{k+1}q_k + q_{k-1};$$

hence we can write

$$\begin{aligned} p_{k+1}q_k - p_kq_{k+1} &= (a_{k+1}p_k + p_{k-1})q_k - p_k(a_{k+1}q_k + q_{k-1}) \\ &= a_{k+1}p_kq_k + p_{k-1}q_k - a_{k+1}p_kq_k - p_kq_{k-1} \qquad (1.34) \\ &= (-1)(p_kq_{k-1} - p_{k-1}q_k). \end{aligned}$$

We assume that the theorem holds for $i = k$, that is, that

$$p_kq_{k-1} - p_{k-1}q_k = (-1)^k.$$

Substituting this result into the last line in (1.34), we see that

$$p_{k+1}q_k - p_kq_{k+1} = (-1)(-1)^k = (-1)^{k+1}.$$

But this is the statement of the theorem for $i = k + 1$, so we have proved that *the theorem holds for* $i = k + 1$ *if it holds for* $i = k$.

We *know* the theorem holds for $i=0$; hence it holds for $i=0+1=1$, and therefore for $i = 1 + 1 = 2$, and so on for all values of $i = 0, 1, 2, \cdots, n$.

Corollary 1.5. *Every convergent* $c_i = p_i/q_i$, $i \geq 1$, *of a simple continued fraction is in its lowest terms, that is,* p_i *and* q_i *have no common divisors other than* $+1$ *or* -1.

Proof. Since

$$p_iq_{i-1} - p_{i-1}q_i = (-1)^i,$$

it follows that any number which divides both p_i and q_i must be a divisor of $(-1)^i$. But the only divisors of $(-1)^i$ are $+1$ and -1; hence the numbers $+1$ and -1 are the only common divisors of p_i and q_i. In our discussion of Euclid's algorithm, we used the symbol $d = (a,b)$ to indicate that d was the g.c.d. of a and b; we can now state that $(p_i,q_i) = 1$, since 1 is the largest number that divides both p_i and q_i.

Problem Set 4

1. Check Theorem 1.4 using the continued fraction [3, 1, 2, 2, 1, 5] by calculating in turn $p_0q_{-1} - p_{-1}q_0$, $p_1q_0 - p_0q_1$, $p_2q_1 - p_1q_2$, etc. Also verify that each convergent p_1/q_1, p_2/q_2, $\cdots$, p_6/q_6 is a rational fraction in its lowest terms.

2. Give another proof of Theorem 1.4 using the following hints. Notice that

$$\begin{aligned} p_iq_{i-1} - p_{i-1}q_i &= (a_ip_{i-1} + p_{i-2})q_{i-1} - p_{i-1}(a_iq_{i-1} + q_{i-2}) \\ &= (-1)(p_{i-1}q_{i-2} - p_{i-2}q_{i-1}). \end{aligned}$$

The expression $p_{i-1}q_{i-2} - p_{i-2}q_{i-1}$ is the same as $p_iq_{i-1} - p_{i-1}q_i$ but with i replaced by $i - 1$. Hence this *reduction*, or "stepdown" from i to $i - 1$, can be repeated, yielding

$$p_{i-1}q_{i-2} - p_{i-2}q_{i-1} = (-1)(p_{i-2}q_{i-3} - p_{i-3}q_{i-2}).$$

After i reductions of the same sort, performed in succession, we obtain the final result,

$$p_iq_{i-1} - p_{i-1}q_i = (-1)^i(p_0q_{-1} - p_{-1}q_0) = (-1)^i \cdot 1 = (-1)^i.$$

1.7 Some Historical Comments

We end this chapter with a few brief remarks concerning the history of the theory of continued fractions. The earliest traces of the idea of a continued fraction are somewhat confused, for many ancient arithmetical results are suggestive of these fractions, but there was no systematic development of the subjec

We have already seen that Euclid's method for finding the g.c.d. of two numbers is essentially that of converting a fraction into a continued fraction. This is perhaps the earliest (c. 300 B.C.) important step in the development of the concept of a continued fraction.

A reference to continued fractions is found in the works of the Indian mathematician Āryabhata, who died around 550 A.D. His work contains one of the earliest attempts at the general solution of a linear indeterminate equation (see next chapter) by the use of continued fractions. Further traces of the general concept of a continued fraction are found occasionally in Arab and Greek writings.

Most authorities agree that the modern theory of continued fractions began with the writings of Rafael Bombelli (born c. 1530), a native of Bologna. His treatise on algebra (1572) contains a chapter on square roots. In our modern symbolism he showed, for example, that

$$\sqrt{13} = 3 + \frac{4}{6+}\,\frac{4}{6+}\cdots.$$

This indicates that he knew, essentially, that

$$\sqrt{a^2+b} = a + \frac{b}{2a+}\,\frac{b}{2a+}\cdots.$$

The next writer to consider these fractions was Pietro Antonio Cataldi (1548–1626), also a native of Bologna. In a treatise on the theory of roots (1613), he expressed $\sqrt{18}$ in the form

$$4.+\cfrac{2}{8.\;\&\cfrac{2}{8.\;\&\cfrac{2}{8.}}}\;.$$

This he modified, for convenience in printing, into the form

$$4.\;\&\frac{2}{8.}\;\&\frac{2}{8.}\;\&\frac{2}{8.},$$

which is substantially the modern form

$$\sqrt{18} = 4 + \frac{2}{8+}\,\frac{2}{8+}\,\frac{2}{8+}\cdots.$$

A third early writer who deserves mention is Daniel Schwenter (1585–1636), who was at various times professor of Hebrew, Oriental languages, and mathematics at the University of Altdorf, Germany. In his book *Geometrica Practica* he found approximations to $\frac{177}{233}$ by finding the g.c.d. of 177 and 233, and from these calculations he determined the convergents $\frac{79}{104}$, $\frac{19}{25}$, $\frac{3}{4}$, $\frac{1}{1}$, and $\frac{0}{1}$.

The next writer of prominence to use continued fractions was Lord Brouncker (1620–1684), the first President of the Royal Society. He transformed the interesting infinite product

$$\frac{4}{\pi} = \frac{3 \cdot 3 \cdot 5 \cdot 5 \cdot 7 \cdot 7 \cdot 9 \cdot 9 \cdot \cdots}{2 \cdot 4 \cdot 4 \cdot 6 \cdot 6 \cdot 8 \cdot 8 \cdot 10 \cdot \cdots},$$

discovered by the English mathematician John Wallis (1655), into the continued fraction

$$\frac{4}{\pi} = 1 + \frac{1^2}{2 +} \frac{3^2}{2 +} \frac{5^2}{2 +} \frac{7^2}{2 + \cdots},$$

but made no further use of these fractions.

In the discussion of Brouncker's fraction in his book *Arithmetica Infinitorum*, published in 1655, Wallis stated a good many of the elementary properties of the convergents to general continued fractions, including the rule for their formation. He also used for the first time the name "continued fraction".

The great Dutch mathematician, mechanician, astronomer, and physicist, Christiaan Huygens (1629–1695) used continued fractions for the purpose of approximating the correct design for the toothed wheels of a planetarium (1698). This is described in his treatise *Descriptio Automati Planetarii*, published posthumously in 1698.

From this beginning great mathematicians such as Euler (1707–1783), Lambert (1728–1777), Lagrange (1736–1813), and many others developed the theory as we know it today. In particular, Euler's great memoir, *De Fractionibus Continiius* (1737), laid the foundation for the modern theory.

Continued fractions play an important role in present day mathematics. They constitute a most important tool for new discoveries in the theory of numbers and in the field of Diophantine approximations. There is the important generalization of continued fractions called the analytic theory of continued fractions, an extensive area for present and future research. In the computer field, continued fractions are used to give approximations to various complicated functions, and once coded for the electronic machines, give rapid numerical results valuable to scientists and to those working in applied mathematical fields.†

† See F. B. Hildebrand, *Introduction to Numerical Analysis*, New York: McGraw-Hill Book Company, 1956 (Chapter 9).

CHAPTER TWO

Diophantine Equations

2.1 Introduction

A great many puzzles, riddles, and trick questions lead to mathematical equations whose solutions must be integers. Here is a typical example: A farmer bought a number of cows at \$80 each, and a number of pigs at \$50 each. His bill was \$810. How many cows and how many pigs did he buy?

If x is the number of cows and y the number of pigs, we have the equation

$$80x + 50y = 810 \tag{2.1}$$

which is equivalent to

$$8x + 5y = 81. \tag{2.2}$$

If nothing limits the values of x and y in equation (2.2), we can give x *any* value, say $x = \frac{1}{2}$, and then solve the resulting equation

$$4 + 5y = 81$$

for y, getting $y = \frac{77}{5}$. In this sense, (2.2) is an *indeterminate* equation, which means that we can always find some value of y corresponding to *any* value we choose for x.

If, however, we restrict the values of x and y to be integers, as the farmer is likely to do (since he is probably not interested in half a cow), then our example belongs to an extensive class of problems requiring that we search for integral solutions x and y of indeterminate equations. Indeterminate equations to be solved in integers (and sometimes in rational numbers) are often called *Diophantine equations* in honor of Diophantus, a Greek mathematician of about the third century A.D., who wrote a book about such equations. Our problem, it should be noted, has the *further restriction* that both x and y must not only be integers but must be positive.

Equation (2.2) and hence equation (2.1) can be solved in many ways. *In fact there is no harm in solving such equations by trial and error or by making intelligent guesses.* For example, if we write equation (2.2) in the form

$$81 - 8x = 5y,$$

we need only search for positive integral values of x such that $81 - 8x$ is a multiple of 5. Letting x, in turn, take on the values $0, 1, 2, 3, \cdots, 10$, we find that $x = 2$ and $x = 7$ are the only non-negative values which make $81 - 8x$ a non-negative multiple of 5. The calculations are

$$x = 2, \qquad 81 - 8x = 81 - 16 = 65 = 5 \cdot 13 = 5y, \qquad y = 13,$$

$$x = 7, \qquad 81 - 8x = 81 - 56 = 25 = 5 \cdot 5 \;\; = 5y, \qquad y = 5;$$

hence the two solutions to our problem are $(x, y) = (2, 13)$ and $(x, y) = (7, 5)$. So the farmer could buy 2 cows and 13 pigs, or 7 cows and 5 pigs.

There are other ways of solving Diophantine equations. We shall give two additional methods. The first of these was used extensively by Euler in his popular text *Algebra*, published in 1770. The second method will show how the theory of continued fractions can be applied to solve such equations.

2.2 The Method Used Extensively by Euler†

Let us consider again the equation

$$8x + 5y = 81. \tag{2.3}$$

Since y has the smaller coefficient, we solve the equation for y, getting

$$y = \frac{81 - 8x}{5}. \tag{2.4}$$

† For additional examples, see O. Ore [10].

Both 81 and 8 contain multiples of 5, that is,

$$81 = 5 \cdot 16 + 1 \qquad \text{and} \qquad 8 = 5 \cdot 1 + 3;$$

therefore, from (2.4), we have

$$y = \frac{(5 \cdot 16 + 1) - (5 \cdot 1 + 3)x}{5}$$

$$= (16 - x) + \frac{1 - 3x}{5} \tag{2.5}$$

$$= (16 - x) + t,$$

where

$$t = \frac{1 - 3x}{5},$$

or

$$3x + 5t = 1. \tag{2.6}$$

Since x and y must be integers, we conclude from equation (2.5) that t must be an integer. Our task, therefore, is to find integers x and t satisfying equation (2.6). This is the essential idea in Euler's method, i.e., to show that integral solutions of the given equation are in turn connected with integral solutions of similar equations with smaller coefficients.

We now reduce this last equation to a simpler one exactly as we reduced (2.3) to (2.6). Solving (2.6) for x, the term with the smaller coefficient, we get

$$x = \frac{1 - 5t}{3} = \frac{1 - (2 \cdot 3 - 1)t}{3}$$

$$= -2t + \frac{t + 1}{3} \tag{2.7}$$

$$= -2t + u,$$

where

$$u = \frac{t + 1}{3},$$

or

$$t = 3u - 1. \tag{2.8}$$

Again, since x and t must be integers, u must also be an integer.

Conversely, if u is an integer, equation (2.8) shows that

$$t = 3u - 1$$

is an integer; x also is an integer since, from (2.7),

$$x = -2t + u = -2(3u - 1) + u = 2 - 5u.$$

Substituting $x = 2 - 5u$ and $t = 3u - 1$ in (2.5) gives

$$y = 16 - 2 + 5u + 3u - 1 = 8u + 13,$$

so that y is an integer. This shows that the general integral solution of (2.3) is

$$\begin{aligned} x &= 2 - 5u, \\ y &= 13 + 8u, \end{aligned} \tag{2.9}$$

where u is any integer, positive, negative, or zero, i.e.,

$$u = 0, \pm 1, \pm 2, \pm 3, \cdots .$$

A direct substitution into (2.3) shows indeed that

$$8x + 5y = 8(2 - 5u) + 5(13 + 8u) = 81.$$

Consequently (2.3) has an infinite number of solutions, one for each integral value of u. A few solutions are listed below:

u	-2	-1	0	1	2	3
x	12	7	2	-3	-8	-13
y	-3	5	13	21	29	37

If the problem is such that we are limited to positive values of x and y, then two *inequalities* must be solved. For example, if in (2.9) both x and y are to be positive, we must solve the two inequalities

$$2 - 5u > 0, \qquad 13 + 8u > 0,$$

for u. These inequalities require that u be an integer such that

$$u < \frac{2}{5}, \qquad \text{and} \qquad u > -\frac{13}{8},$$

and a glance at Figure 1 shows that the only two possible integral values of u are 0 and -1. Substituting, in turn, $u = 0$ and $u = -1$ in (2.9) gives $(x,y) = (2,13)$ and $(x,y) = (7,5)$, the original answers to the farmer's problem.

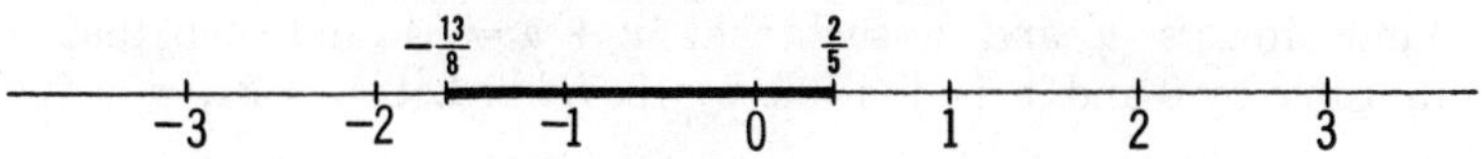

Figure 1

Going back over the solution of equation (2.3) we can raise certain questions. For example, why should we solve for y, rather than for x, simply because y has the smaller coefficient? If we had solved first for x, could we have arrived at a shorter solution? In the second line below equation (2.4) we replaced 8 by $5 \cdot 1 + 3$. Why not replace 8 by $5 \cdot 2 - 2$? In solving equation (2.3) the writer did not have in mind the presentation of the shortest solution. We leave it to the reader to experiment and try to obtain general solutions in the least number of steps.

Problem Set 5

1. Use Euler's method to solve the following linear Diophantine equations. In each case list the positive integral solutions, if any.

   (a) $15x + 47y = 2$ (c) $15x + 47y = 4$
   (b) $31x + 7y = 1$ (d) $13x + 21y = 295$

2. Does the indeterminate equation $6x + 15y = 17$ have integral solutions? Note that the left side of the equation is divisible by 3. What about the right-hand side? What happens if we go ahead and use Euler's method anyway?

3. Return to equation (2.9) and fill out the following table for the values of u indicated.

u	-3	-2	-1	0	1	2	3	4
x								
y								

   On ordinary graph paper plot the points (x, y) and join them by a straight line. Use this graph to pick out the positive solutions of the equation $8x + 5y = 81$.

4. A man buys horses and cows for a total amount of \$2370. If one horse costs \$37 and one cow \$22, how many horses and cows does he buy?

5. Show that the equation $17x - 15y = 5$ has infinitely many positive integral solutions.

6. Find integers u and v such that $u + v = 84$ and such that u is divisible by 9 and v is divisible by 13. *Hint:* Let $u = 9x$, $v = 13y$.

7. Find a number N which leaves a remainder 2 when divided by 20 and a remainder 12 when divided by 30. *Hint:* Find integers x and y so that the required number $N = 20x + 2 = 30y + 12$. Hence solve the equation $20x - 30y = 10$.

2.3 The Indeterminate Equation $ax - by = \pm 1$

We are now ready to show how continued fractions can be used to solve the linear indeterminate equation $ax + by = c$ where a, b, and c are given integers, and where x and y are the unknown integers.

Our approach to this will be a step-by-step process, through easy stages, culminating in the final mastery of the solution of *any solvable* equation of the form $ax + by = c$. We start with the restrictions that the coefficients of x and y are of different signs and that they have no common divisor but 1. Thus we first learn to solve the equation

$$ax - by = 1, \qquad (a, b) = 1, \tag{2.10}$$

where a and b are positive integers. [The equation $-ax + by = 1$, $(a, b) = 1$, is of the same form with the roles of x and y interchanged.] The integers a and b can have *no divisors greater than* 1 *in common*; for, if an integer d divides both a and b, it also divides the integer 1 on the right-hand side of the equation and hence can have only the value $d = 1$. In other words, a and b must be relatively prime, or $d = (a, b) = 1$.

We shall now state and prove

THEOREM 2.1. *The equation $ax - by = 1$, where a and b are relatively prime positive integers, has an infinite number of integral solutions (x, y).*

We first convert a/b into a finite simple continued fraction

$$\frac{a}{b} = [a_1, a_2, \cdots, a_{n-1}, a_n], \tag{2.11}$$

and calculate the convergents $c_1, c_2, \cdots, c_{n-1}, c_n$. The last two convergents,

$$c_{n-1} = \frac{p_{n-1}}{q_{n-1}}, \qquad c_n = \frac{p_n}{q_n} = \frac{a}{b},$$

are the key to the solution, for they satisfy the relation stated in Theorem 1.4, namely that

$$p_n q_{n-1} - q_n p_{n-1} = (-1)^n,$$

and since $p_n = a$, $q_n = b$, this gives

$$aq_{n-1} - bp_{n-1} = (-1)^n. \tag{2.12}$$

If n is *even*, that is if we have an even number of partial quotients $a_1, a_2, \cdots, a_n$, then $(-1)^n = 1$ and (2.12) becomes

$$aq_{n-1} - bp_{n-1} = 1. \tag{2.13}$$

Comparing this with the given equation

$$ax - by = 1,$$

we see that a solution to this equation is

$$x_0 = q_{n-1}, \qquad y_0 = p_{n-1}.$$

This, however, is a *particular solution* and not the *general solution* We indicate particular solutions by the notation (x_0, y_0).

On the other hand, if n is *odd* so that $(-1)^n = -1$, we can modify the continued fraction expansion (2.11) by replacing

$$\frac{1}{a_n} \quad \text{by} \quad \frac{1}{(a_n - 1) + \frac{1}{1}} \qquad \text{if } a_n > 1,$$

or by replacing

$$\frac{1}{a_{n-1} + \frac{1}{a_n}} \quad \text{by} \quad \frac{1}{a_{n-1} + 1} \qquad \text{if } a_n = 1.$$

Thus, if (2.11) has an odd number of partial quotients, it may be transformed into

$$[a_1, a_2, \cdots, a_n - 1, 1], \qquad \text{if } a_n > 1,$$

or into

$$[a_1, a_2, \cdots, a_{n-1} + 1], \qquad \text{if } a_n = \;,$$

in both cases the number of partial quotients is even. Using these continued fractions, one case or the other, we *re-calculate* p_{n-1}/q_{n-1} and $p_n/q_n = a/b$, and equation (2.13) is satisfied once more.

Once a particular solution (x_0, y_0) of equation (2.10) has been found, it is an easy matter to find the *general solution*. To this end, let (x, y) be *any* other solution of (2.10). Then

$$ax - by = 1,$$

and

$$ax_0 - by_0 = 1,$$

and a subtraction gives

$$a(x - x_0) = b(y - y_0). \tag{2.14}$$

This shows that b divides the left side of the equation. But b cannot divide a since a and b are relatively prime; hence b must divide $x - x_0$, that is, $x - x_0$ is an integral multiple of b, and we may write

$$x - x_0 = tb \qquad (t \text{ an integer}),$$

or

$$x = x_0 + tb.$$

But if this is true, (2.14) shows that

$$a(tb) = b(y - y_0),$$

so that

$$y - y_0 = at.$$

It follows that any other solution (x,y) of $ax - by = 1$ has the form

$$\begin{aligned} x &= x_0 + tb \\ y &= y_0 + ta \end{aligned} \qquad t = 0, \pm 1, \pm 2, \pm 3, \cdots. \tag{2.15}$$

Conversely, if (x_0, y_0) is any particular solution of $ax - by = 1$, and if we set up the equations (2.15) with t any integer whatever, then the values (x, y) will satisfy the given equation, because

$$\begin{aligned} ax - by &= a(x_0 + tb) - b(y_0 + ta) \\ &= (ax_0 - by_0) + tab - tab \\ &= ax_0 - by_0 = 1. \end{aligned}$$

We call the values of x and y given by equations (2.15) the *general solution* of the indeterminate equation $ax - by = 1$.

EXAMPLE 1. Find integral solutions of the indeterminate equation

$$205x - 93y = 1.$$

Here the integers $205 = 5 \cdot 41$ and $93 = 3 \cdot 31$ are relatively prime, so the equation has solutions.

SOLUTION. The continued fraction $\frac{205}{93} = [2, 4, 1, 8, 2]$ has an *odd* number of partial quotients, but it can be replaced by

$$\frac{205}{93} = [2, 4, 1, 8, 1, 1],$$

the equivalent expansion with an *even* number of quotients. The convergents are then computed.

i	-1	0	1	2	3	4	5	6
a_i			2	4	1	8	1	1
p_i	0	1	2	9	11	97	108	205
q_i	1	0	1	4	5	44	49	93
c_i			$\frac{2}{1}$	$\frac{9}{4}$	$\frac{11}{5}$	$\frac{97}{44}$	$\frac{108}{49}$	$\frac{205}{93}$

Here $n = 6$, $p_{n-1} = p_5 = 108 = y_0$, $q_{n-1} = q_5 = 49 = x_0$, and hence, by (2.15), the general solution of the equation $ax - by = 205x - 93y = 1$ is

$$\begin{aligned} x &= x_0 + tb = 49 + 93t \\ y &= y_0 + ta = 108 + 205t \end{aligned} \qquad t = 0, \pm 1, \pm 2, \cdots .$$

As a check, let $t = 1$; then $x = 142$, $y = 313$ and $205(142) - 93(313) = 29110 - 29109 = 1$. As a general check we have

$$205(49 + 93t) - 93(108 + 205t) = 1,$$

since the terms involving t cancel.

The method for solving the equation

$$ax - by = -1, \qquad (a,b) = 1,$$

is quite similar to that used to solve (2.10). We convert a/b into a finite simple continued fraction with an *odd* number of convergents. In this case equation (2.12) becomes

$$aq_{n-1} - bp_{n-1} = (-1)^n = -1,$$

since n is odd. Comparing this equation with

$$ax - by = -1,$$

we see that

$$x_0 = q_{n-1}, \qquad y_0 = p_{n-1}$$

is a *particular solution* of the given equation, the *general solution* being, as before,

$$\begin{aligned} x &= x_0 + tb \\ y &= y_0 + ta \end{aligned} \qquad t = 0, \pm 1, \pm 2, \pm 3, \cdots .$$

EXAMPLE 2. Find integral solutions of the equation

$$205x - 93y = -1.$$

SOLUTION. The numbers 205 and 93 are relatively prime, hence the given equation has integral solutions. The continued fraction expansion for $\frac{205}{93}$ is

$$\frac{205}{93} = [2, 4, 1, 8, 2]$$

and has an odd number of partial quotients, so $(-1)^n = (-1)^5 = -1$ as required. To find the convergents we set up the table

i	-1	0	1	2	3	4	5
a_i			2	4	1	8	2
p_i	0	1	2	9	11	97	205
q_i	1	0	1	4	5	44	93
c_i			$\frac{2}{1}$	$\frac{9}{4}$	$\frac{11}{5}$	$\frac{97}{44}$	$\frac{205}{93}$

Our calculations show that $c_{n-1} = p_{n-1}/q_{n-1} = p_4/q_4 = \frac{97}{44}$; hence a particular solution of the given equation is $x_0 = q_4 = 44$ and $y_0 = p_4 = 97$. The general solution, therefore, is

$$\begin{aligned} x &= x_0 + tb = 44 + 93t \\ y &= y_0 + ta = 97 + 205t \end{aligned} \qquad t = 0, \pm 1, \pm 2, \cdots .$$

As a check, take $t = -1$; then $(x,y) = (-49, -108)$, and

$$205(-49) - 93(-108) = -10045 + 10044 = -1.$$

It is interesting to notice that once we have calculated the particular solution $(x_0, y_0) = (q_{n-1}, p_{n-1})$ of the equation

$$ax - by = 1,$$

we can immediately obtain a particular solution, call it (x_1, y_1), of the equation

$$ax - by = -1. \tag{2.16}$$

The particular solution of (2.16) will be

$$\begin{aligned} x_1 &= b - x_0 = b - q_{n-1}, \\ y_1 &= a - y_0 = a - p_{n-1}, \end{aligned} \tag{2.17}$$

for then

$$\begin{aligned} ax_1 - by_1 &= a(b - q_{n-1}) - b(a - p_{n-1}) \\ &= ab - ab - (aq_{n-1} - bp_{n-1}) \\ &= (-1)(aq_{n-1} - bp_{n-1}) \\ &= (-1)(+1) = -1, \end{aligned}$$

since from (2.13) we know that $aq_{n-1} - bp_{n-1} = 1$.

The general solution of the equation $ax - by = -1$ will then be

$$\begin{aligned} x &= x_1 + tb \\ y &= y_1 + ta \end{aligned} \qquad t = 0, \pm 1, \pm 2, \pm 3, \cdots, \tag{2.18}$$

and this can be checked by a direct substitution.

Example 3. Show that we can solve Example 2 if we have already solved Example 1. That is, solve the equation $205x - 93y = -1$, knowing that $(x_0, y_0) = (49, 108)$ is a particular solution of $205x - 93y = +1$.

Solution. Using equations (2.17) we find that

$$\begin{aligned} x_1 &= b - x_0 = 93 - 49 = 44, \\ y_1 &= a - y_0 = 205 - 108 = 97, \end{aligned}$$

is a particular solution of $205x - 93y = -1$. Hence the general solution, according to (2.18), is

$$\begin{aligned} x &= 44 + 93t \\ y &= 97 + 205t \end{aligned} \qquad t = 0, \pm 1, \pm 2, \cdots, \tag{2.19}$$

which agrees with the solution given for Example 2.

There is still another way to solve Example 2, provided we know a particular solution of Example 1. This is illustrated in the following example.

EXAMPLE 4. Give a third solution of the equation $205x - 93y = -1$.

SOLUTION. Since $(x_0, y_0) = (49, 108)$ is a particular solution of the equation $205x - 93y = +1$, we know that

$$205(49) - 93(108) = +1.$$

If we multiply through by -1 we see that

$$205(-49) - 93(-108) = -1;$$

hence $(x_1, y_1) = (-49, -108)$ is a particular solution of $205x - 93y = -1$, and the general solution becomes

$$\begin{aligned} x &= x_1 + tb = -49 + 93t \\ y &= y_1 + ta = -108 + 205t \end{aligned} \qquad t = 0, \pm 1, \pm 2, \cdots. \tag{2.20}$$

Notice that equations (2.19) and (2.20) reproduce the same values of x and y *but not for the same values of* t. For example, $t = 2$ in (2.19) gives $(x, y) = (230, 507)$, the same values obtained from (2.20) for $t = 3$.

Problem Set 6

1. Find the general integral solutions of the following equations. Check each answer.

   (a) $13x - 17y = 1$
   (b) $13x - 17y = -1$
   (c) $65x - 56y = 1$
   (d) $65x - 56y = -1$
   (e) $56x - 65y = 1$

2.4 The General Solution of $ax - by = c$, $(a, b) = 1$

Once we have learned to solve the indeterminate equation

$$ax - by = 1, \tag{2.21}$$

where a and b are two relatively prime positive integers, it is a simple matter to solve the equation

$$ax - by = c, \tag{2.22}$$

where c is any integer. For, suppose that (x_0, y_0) is any particular solution of (2.21); then

$$ax_0 - by_0 = 1,$$

and multiplying both sides by c, we obtain

$$a(cx_0) - b(cy_0) = c,$$

so that (cx_0, cy_0) is a particular solution of (2.22). Thus the general solution of equation (2.22) will be

$$\begin{aligned} x &= cx_0 + bt \\ y &= cy_0 + at \end{aligned} \qquad t = 0, \pm 1, \pm 2, \cdots. \tag{2.23}$$

This can easily be verified by a direct substitution into (2.22).

EXAMPLE 1. Solve the equation

$$205x - 93y = 5.$$

SOLUTION. From Example 1, Section 2.3, we know that $(x_0, y_0) = (49, 108)$ is a particular solution of the equation $205x - 93y = 1$, that is,

$$205(49) - 93(108) = 1.$$

Multiplying both sides by 5 we get

$$205(5 \cdot 49) - 93(5 \cdot 108) = 5,$$

so that $(5x_0, 5y_0) = (245, 540)$ is a particular solution of the given equation. The general solution, according to (2.23), will be

$$\begin{aligned} x &= 245 + 93t \\ y &= 540 + 205t \end{aligned} \qquad t = 0, \pm 1, \pm 2, \cdots.$$

As a check, take $t = 1$; then $(x, y) = (338, 745)$ and

$$205(338) - 93(745) = 69290 - 69285 = 5.$$

EXAMPLE 2. Solve the equation

$$205x - 93y = -5.$$

SOLUTION. In Example 1 of this section we recalled that

$$205(49) - 93(108) = 1.$$

Multiplying through by -5 we get

$$205(-5 \cdot 49) - 93(-5 \cdot 108) = -5,$$

or

$$205(-245) - 93(-540) = -5,$$

so that $(x_0, y_0) = (-245, -540)$ is a particular solution of the given equation. The general solution, according to equation (2.23), is then

$$\begin{aligned} x &= -245 + 93t \\ y &= -540 + 205t \end{aligned} \qquad t = 0, \pm 1, \pm 2, \cdots .$$

To check this, take $t = 2$, then $(x, y) = (-59, -130)$, and

$$205(-59) - 93(-130) = -12095 + 12090 = -5.$$

Problem Set 7

1. Use particular solutions obtained from the problems at the end of Section 2.3 to obtain the general integral solutions of the following equations. Check each answer.

   (a) $13x - 17y = 5$ (b) $65x - 56y = 7$ (c) $56x - 65y = -3$

2.5 The General Solution of $ax + by = c$, $(a, b) = 1$

The discussion of this equation is similar, except for some minor changes, to that of the equation $ax - by = c$. Still assuming that a and b are positive integers, we first find a particular solution of the equation

$$ax + by = 1, \qquad (a,b) = 1.$$

To do this, expand a/b as a simple continued fraction with an *even* number of partial quotients. From the table of convergents read off p_{n-1} and q_{n-1}. Then

$$aq_{n-1} - bp_{n-1} = 1,$$

as before. The trick now is to write the given equation $ax + by = c$ in the form

$$ax + by = c \cdot 1 = c(aq_{n-1} - bp_{n-1}).$$

Rearrange terms to obtain

$$a(cq_{n-1} - x) = b(y + cp_{n-1}). \tag{2.24}$$

This shows that b divides the left side of the equation; but $(a, b) = 1$, so b cannot divide a. Therefore b divides $cq_{n-1} - x$, so that there is an integer t such that

$$cq_{n-1} - x = tb, \tag{2.25}$$

or

$$x = cq_{n-1} - tb \tag{2.26}$$

Substitute (2.25) into (2.24) to get

$$a(tb) = b(y + cp_{n-1}),$$

and solve for y to obtain

$$y = at - cp_{n-1}. \tag{2.27}$$

Conversely, for any integer t, a direct substitution of (2.26) and (2.27) into $ax + by$ gives

$$\begin{aligned} ax + by &= a(cq_{n-1} - tb) + b(at - cp_{n-1}) \\ &= acq_{n-1} - tab + tab - bcp_{n-1} \\ &= c(aq_{n-1} - bp_{n-1}) = c \cdot 1 = c, \end{aligned}$$

so the equation $ax + by = c$ is satisfied. Thus the general solution of the equation $ax + by = c$ is

$$\begin{aligned} x &= cq_{n-1} - tb \\ y &= at - cp_{n-1} \end{aligned} \qquad t = 0, \pm 1, \pm 2, \pm 3, \cdots. \tag{2.28}$$

Example 1. Solve the indeterminate equation

$$13x + 17y = 300.$$

Solution. We find that $(x_0, y_0) = (4, 3)$ is a particular solution of the equation

$$13x - 17y = 1,$$

or that $13(4) - 17(3) = 1$, and so the given equation may be written in the form

$$13x + 17y = 300(13 \cdot 4 - 17 \cdot 3),$$

or

$$13x - 13(4 \cdot 300) = -17y - 17(3 \cdot 300).$$

This shows that

$$13(x - 1200) = -17(y + 900), \tag{2.29}$$

so that 17 divides $x - 1200$, or

$$x = 1200 + 17t.$$

Replacing $x - 1200$ by $17t$ in (2.29) gives

$$y = -13t - 900.$$

Hence the general solution of the given equation is

$$\begin{aligned} x &= 1200 + 17t \\ y &= -13t - 900 \end{aligned} \qquad t = 0, \pm 1, \pm 2, \pm 3, \cdots.$$

EXAMPLE 2. Solve the indeterminate equation

$$13x + 17y = -300.$$

SOLUTION. The second equation in the solution of Example 1 now becomes

$$13x + 17y = -300(13 \cdot 4 - 17 \cdot 3),$$

and equation (2.29) is replaced by

$$13(x + 1200) = -17(y - 900). \tag{2.29a}$$

It follows that 17 divides $x + 1200$, or

$$x = -1200 + 17t,$$

and replacing $x + 1200$ by $17t$ gives

$$y = 900 - 13t.$$

Hence the general solution of the given equation is

$$\begin{aligned} x &= -1200 + 17t \\ y &= 900 - 13t \end{aligned} \qquad t = 0, \pm 1, \pm 2, \pm 3, \cdots.$$

2.6 The General Solution of $Ax \pm By = \pm C$

By multiplying through by -1, any equation of the form

$$\pm Ax \pm By = C$$

can be reduced to one or the other of the forms

$$Ax + By = \pm C, \qquad Ax - By = \pm C, \tag{2.30}$$

where A and B are positive integers. For example, of the four equations

$$3x + 7y = 10, \quad 3x - 7y = 10, \quad -3x - 7y = 10, \quad -3x + 7y = 10,$$

the first two are already in the required form, and the second two can be replaced, respectively, by

$$3x + 7y = -10 \qquad \text{and} \qquad 3x - 7y = -10.$$

Not all equations of the form (2.30) have solutions. To see this, let d be the greatest common divisor of A and B. Then, if d *does not divide* C, neither of the equations (2.30) can be solved in integers x, y, for the left side of each would be divisible by d while the right side is not.

On the other hand, if d does divide C, then we can divide both sides of the equations (2.30) by d, reducing them respectively to equations of the form we have just discussed, namely

$$(2.31) \qquad ax + by = c, \qquad ax - by = c,$$

where a and b are relatively prime, and of which we know the solutions. Conversely, any solution of equations (2.31) will automatically give solutions of equations (2.30).

EXAMPLE 1. Solve the equation

$$410x - 186y = 10.$$

SOLUTION. Since $410 = 2 \cdot 5 \cdot 41$, $186 = 2 \cdot 3 \cdot 31$, the g.c.d. of 410 and 186 is $d = 2$. Since $d = 2$ divides 10, the equation can be solved. Divide the given equation by 2 to obtain

$$205x - 93y = 5,$$

where now 205 and 93 are relatively prime. This is the equation solved in Example 1 of Section 2.4. The general solution of $205x - 93y = 5$ found there was

$$x = 245 + 93t,$$

$$y = 540 + 205t,$$

and substituting it into $410x - 186y$ we find that

$$410(245 + 93t) - 186(540 + 205t) = 410 \cdot 245 - 186 \cdot 540 = 10.$$

The main results obtained from our study of the linear Diophantine equation can be summarized as follows:

Summary. Any equation of the form $Ax \pm By = \pm C$ has integral solutions x,y only if the greatest common divisor of A and B divides C. In this case, divide A, B, and C by $d = (A, B)$, reducing the given equation to either the form

$$\text{(i)} \qquad ax + by = c,$$

or the form

$$\text{(ii)} \qquad ax - by = c,$$

where in both equations a and b are relatively prime positive integers, and where c is a positive or negative integer. The next step is to expand a/b as a simple continued fraction with an *even* number

n of partial quotients, and from the table of convergents read off p_{n-1} and q_{n-1}. Then $aq_{n-1} - bp_{n-1} = 1$, and the general solution of (i) is

$$\text{(iii)}\qquad \begin{aligned} x &= cq_{n-1} - tb \\ y &= ta - cp_{n-1} \end{aligned} \qquad t = 0, \pm 1, \pm 2, \cdots .$$

Likewise the general solution of (ii) is

$$\text{(iv)}\qquad \begin{aligned} x &= cq_{n-1} + tb \\ y &= cp_{n-1} + ta \end{aligned} \qquad t = 0, \pm 1, \pm 2, \cdots .$$

The solutions (iii) and (iv) represent, respectively, for the cases (i) and (ii) the general solution of $Ax \pm By = \pm C$.

Problem Set 8

1. Two of these six equations do not have integral solutions. Find the general solution in integers of the others.

(a) $183x + 174y = 9$	(d) $34x - 49y = 5$
(b) $183x - 174y = 9$	(e) $34x + 49y = 5$
(c) $77x + 63y = 40$	(f) $56x + 20y = 11$

2. Express $\frac{68}{77}$ as the sum of two fractions whose denominators are 7 and 11. *Hint:* Find integers x and y such that $\dfrac{68}{77} = \dfrac{x}{7} + \dfrac{y}{11}$.

3. The sum of two positive integers a and b is 100. If a is divided by 7 the remainder is 5, and if b is divided by 9 the remainder is also 5. Find a and b. *Hint:* Let $a = 7x + 5$, $b = 9y + 5$ and use the fact that $a + b = 100$.

4. Find positive integral solutions (x, y) of $13x + 17y = 300$.

2.7 Sailors, Coconuts, and Monkeys

The following problem is of considerable age and, in one form or another, continues to appear from time to time.

Five sailors were cast away on an island. To provide food, they collected all the coconuts they could find. During the night one of the sailors awoke and decided to take his share of the coconuts. He divided the nuts into five equal piles and discovered that one nut was left over, so he threw this extra one to the monkeys. He then hid his share and went back to sleep. A little later a second sailor awoke and had the same idea as the first. He divided the remainder of the nuts into five equal piles, discovered also that one was left over, and threw it to the monkeys. Then he hid his share. In their turn the other three sailors did the same thing, each throwing a coconut to the monkeys.

The next morning the sailors, all looking as innocent as possible, divided the remaining nuts into five equal piles, no nuts being left over this time. *The problem is to find the smallest number of nuts in the original pile.*

In order to solve this problem, let x be the original number of coconuts. The first sailor took $\frac{1}{5}(x-1)$ coconuts and left $\frac{4}{5}(x-1)$. Similarly the second sailor took

$$\frac{1}{5}\left[\frac{4}{5}(x-1)-1\right]=\frac{4x-9}{25}$$

coconuts and left four times this number, or

$$\frac{16x-36}{25}.$$

Similarly, we find that the third, fourth, and fifth sailors left, respectively,

$$\frac{64x-244}{125}, \qquad \frac{256x-1476}{625}, \qquad \frac{1024x-8404}{3125}$$

nuts.

Now the number of nuts in the last pile must be a multiple of 5 since it was divided evenly into five piles with no nuts left over. Hence

$$\frac{1024x-8404}{3125}=5y,$$

where y is some integer. Multiplying both sides by 3125 we obtain the indeterminate equation

$$(2.32) \qquad 1024x-15625y=8404.$$

Factoring into primes we find $1024=2^{10}$ and $15625=5^6$; hence these numbers are relatively prime and the equation (2.32) has integral solutions. We first seek a particular solution (x_1, y_1) of the equation

$$(2.33) \qquad 1024x-15625y=1.$$

To this end, the convergents of the continued fraction

$$\frac{1024}{15625}=[0, 15, 3, 1, 6, 2, 1, 3, 2, 1]$$

are calculated:

i	-1	0	1	2	3	4	5	6	7	8	9	10
a_i			0	15	3	1	6	2	1	3	2	1
p_i	0	1	0	1	3	4	27	58	85	313	711	1024
q_i	1	0	1	15	46	61	412	885	1297	4776	10849	15625
c_i											$\frac{711}{10849}$	

The convergent c_9 yields the particular solution $x_1 = q_9 = 10849$, $y_1 = p_9 = 711$ of equation (2.33). Hence $x_0 = 8404x_1 = 91174996$, $y_0 = 8404y_1 = 5975244$ will be a particular solution of equation (2.32). The general solution is

$$(2.34)\qquad \begin{aligned} x &= 91174996 + 15625t \\ y &= 5975244 + 1024t \end{aligned} \qquad t = 0, \pm 1, \pm 2, \cdots .$$

Since both x and y must be positive, we search for the value of t which gives the smallest positive value of x and which at the same time makes y positive. From (2.34) we find that t must be an integer satisfying the two inequalities

$$t > -\frac{91174996}{15625} = -5835.2\cdots,$$

$$t > -\frac{5975244}{1024} = -5835.1\cdots.$$

Hence the required value is $t = -5835$. Introducing this value of t into equations (2.34), we finally obtain

$$x = 91174996 - 91171875 = 3121,$$

$$y = 5975244 - 5975040 = 204,$$

which means that the original number of coconuts was 3121 and each sailor received 204 in the final distribution.

For an interesting discussion of this and related problems, see the article entitled "Mathematical Games" by Martin Gardner in *Scientific American*, April, 1958. One should also keep in mind the excellent collection of references, *Recreational Mathematics, A Guide to the Literature*, by William L. Schaaf, published by the National Council of Teachers of Mathematics.

CHAPTER THREE

Expansion of Irrational Numbers

3.1 Introduction

So far our discussion has been limited to the expansion of rational numbers. We proved that a rational number can be expanded into a finite simple continued fraction, and, conversely, every finite simple continued fraction represents a rational number.

This chapter will deal with the simple continued fraction expansion of *irrational* numbers, and we shall see that these fractions do not terminate but go on forever.

An *irrational number* is one which cannot be represented as the ratio of two integers. The numbers

$$\sqrt{2}, \quad \sqrt{3}, \quad 1 \pm \sqrt{2}, \quad \frac{3 \pm \sqrt{7}}{5}$$

are all irrational. Any number of the form

$$\frac{P \pm \sqrt{D}}{Q},$$

where P, D, Q are integers, and where D is a positive integer *not* a perfect square, is irrational. A number of this form is called a *quadratic irrational* or *quadratic surd* since it is the root of the quadratic equation

$$Q^2x^2 - 2PQx + (P^2 - D) = 0.$$

Our discussion will be limited to the expansion of quadratic irrationals.

There are irrational numbers which are not quadratic surds. The irrational number $\pi = 3.14159 \cdots$ is one example. The irrational number $\sqrt{2}$ is the solution of the algebraic equation $x^2 - 2 = 0$, and is therefore called an "algebraic number". An *algebraic number* is a number x which satisfies an algebraic equation, i.e., an equation of the form

$$a_0x^n + a_1x^{n-1} + \cdots + a_n = 0,$$

where $a_0, a_1, \cdots$ are integers, not all zero. A number which is *not* algebraic is called a *transcendental number*. It can be proved that π is transcendental, but this not easy to do.† The number e is also transcendental. It is quite difficult to expand transcendental numbers into continued fractions; using decimal approximations to these numbers, such as $\pi = 3.14159 \cdots$ and $e = 2.71828 \cdots$, we can calculate a few of the first terms of their continued fraction expansions, but the methods of obtaining the expansions of π and e given in Appendix II are beyond the scope of this monograph.

Those who wish to learn about the two classes of irrational numbers, namely algebraic irrational numbers and transcendental numbers, and to study the deeper properties of each should read the first monograph in the NML (New Mathematical Library) series: *Numbers: Rational and Irrational*, by Ivan Niven.

3.2 Preliminary Examples

The procedure for expanding an irrational number is fundamentally the same as that used for rational numbers. Let x be the given irrational number. Calculate a_1, the greatest integer less than x, and express x in the form

$$x = a_1 + \frac{1}{x_2}, \qquad 0 < \frac{1}{x_2} < 1,$$

where the number

$$x_2 = \frac{1}{x - a_1} > 1$$

is *irrational;* for, if an integer is subtracted from an irrational number, the result and the reciprocal of the result are irrational.

To continue, calculate a_2, the largest integer less than x_2, and

† See I. Niven [8].

express x_2 in the form

$$x_2 = a_2 + \frac{1}{x_3}, \qquad 0 < \frac{1}{x_3} < 1, \qquad a_2 \geq 1,$$

where, again, the number

$$x_3 = \frac{1}{x_2 - a_2} > 1$$

is irrational.

This calculation may be repeated indefinitely, producing in succession the equations

$$\begin{array}{lll} x = a_1 + \dfrac{1}{x_2}, & x_2 > 1, & \\[2ex] x_2 = a_2 + \dfrac{1}{x_3}, & x_3 > 1, & a_2 \geq 1, \\[2ex] x_3 = a_3 + \dfrac{1}{x_4}, & x_4 > 1, & a_3 \geq 1, \\ \cdots\cdots\cdots\cdots & \cdots\cdots & \cdots \\ x_n = a_n + \dfrac{1}{x_{n+1}}, & x_{n+1} > 1, & a_n \geq 1, \\ \cdots\cdots\cdots\cdots & \cdots\cdots & \cdots, \end{array} \tag{3.1}$$

where $a_1, a_2, \cdots, a_n, \cdots$ are all integers and where the numbers $x, x_2, x_3, x_4, \cdots$ are all irrational. This process cannot terminate, for the only way this could happen would be for some integer a_n to be equal to x_n, which is impossible since each successive x_i is irrational.

Substituting x_2 from the second equation in (3.1) into the first equation, then x_3 from the third into this result, and so on, produces the required infinite simple continued fraction

$$x = a_1 + \frac{1}{x_2} = a_1 + \cfrac{1}{a_2 + \cfrac{1}{x_3}} = a_1 + \cfrac{1}{a_2 + \cfrac{1}{a_3 + \cfrac{1}{x_4}}} = \cdots,$$

or

$$x = [a_1, a_2, a_3, a_4, \cdots],$$

where the three dots indicate that the process is continued indefinitely.

Before discussing some of the more "theoretical" aspects of *infinite simple continued fractions*, an example or two should be worked to make sure the expansion procedure is understood.

EXAMPLE 1. Expand $\sqrt{2}$ into an infinite simple continued fraction.

SOLUTION. The largest integer $< \sqrt{2} = 1.414\cdots$ is $a_1 = 1$, so

$$\sqrt{2} = a_1 + \frac{1}{x_2} = 1 + \frac{1}{x_2}\cdot$$

Solving this equation for x_2, we get

$$x_2 = \frac{1}{\sqrt{2}-1}\cdot\frac{\sqrt{2}+1}{\sqrt{2}+1} = \sqrt{2}+1.$$

Consequently,

$$\sqrt{2} = a_1 + \frac{1}{x_2} = 1 + \frac{1}{\sqrt{2}+1}\cdot$$

The largest integer $< x_2 = \sqrt{2}+1 = 2.414\cdots$ is $a_2 = 2$, so

$$x_2 = a_2 + \frac{1}{x_3} = 2 + \frac{1}{x_3},$$

where

$$x_3 = \frac{1}{x_2 - 2} = \frac{1}{(\sqrt{2}+1)-2} = \frac{1}{\sqrt{2}-1}$$

$$= \frac{1}{\sqrt{2}-1}\cdot\frac{\sqrt{2}+1}{\sqrt{2}+1} = \sqrt{2}+1 > 1.$$

At this stage we know that

$$\sqrt{2} = a_1 + \frac{1}{x_2} = 1 + \cfrac{1}{2+\cfrac{1}{x_3}} = 1 + \cfrac{1}{2+\cfrac{1}{\sqrt{2}+1}}\cdot$$

Since $x_3 = \sqrt{2}+1$ is the same as $x_2 = \sqrt{2}+1$, the calculations of x_4, x_5, $\cdots$ will all produce the same result, namely $\sqrt{2}+1$. Thus all the subsequent partial quotients will be equal to 2 and the infinite expansion of $\sqrt{2}$ will be

$$\sqrt{2} = 1 + \frac{1}{2+}\,\frac{1}{2+}\cdots = [1, 2, 2, 2, \cdots] = [1, \overline{2}].$$

The bar over the 2 on the right indicates that the number 2 is repeated over and over.

Immediately some questions are raised. For example, is it possible to prove that the infinite continued fraction $[1, 2, 2, \cdots] = [1, \overline{2}]$ actually represents the irrational number $\sqrt{2}$? Certainly there is more to this than is evident at first glance, and it will be one of the more difficult questions to be discussed in this chapter. We can, however, give a *formal* answer to this question. A *formal* answer means, roughly speaking, that we go through certain manipulations, but no claim is made that every move is necessarily justified. With this understanding, we write

$$x = 1 + \cfrac{1}{2 + \cfrac{1}{2 + \cfrac{1}{2 + \ddots}}},$$

or

$$x - 1 = \cfrac{1}{2 + \cfrac{1}{2 + \cfrac{1}{2 + \ddots}}};$$

hence

$$x = 1 + (x - 1),$$

or $1 = 1$, which tells us nothing about x. However, using the same idea, we can write

$$x = 1 + \cfrac{1}{2 + \left(\cfrac{1}{2 + \cfrac{1}{2 + \ddots}}\right)}$$

$$= 1 + \frac{1}{2 + (x - 1)} = 1 + \frac{1}{x + 1},$$

from which we see that

$$x - 1 = \frac{1}{x + 1},$$

so

$$(x - 1)(x + 1) = 1, \qquad \text{or} \qquad x^2 = 2.$$

Thus

$$x = 1 + \frac{1}{2+}\,\frac{1}{2+}\cdots = \sqrt{2}.$$

Some additional examples of a similar sort are:

$$\sqrt{3} = [1, 1, 2, 1, 2, 1, 2, \cdots] = [1, \overline{1, 2}],$$

$$\sqrt{15} = [3, 1, 6, 1, 6, \cdots] = [3, \overline{1, 6}],$$

$$\sqrt{31} = [5, \overline{1, 1, 3, 5, 3, 1, 1, 10}].$$

In each of these examples the numbers under the bar form the *periodic part* of the expansion, the number $\sqrt{31}$ having quite a long period. These examples are illustrations of a theorem first proved by Lagrange in 1770 to the effect that *the continued fraction expansion of any quadratic irrational is periodic after a certain stage.* This theorem will be proved in Chapter 4.

EXAMPLE 2. Find the infinite continued fraction expansion for

$$x = \frac{25 + \sqrt{53}}{22}.$$

SOLUTION. We proceed exactly as in Example 1. Since $\sqrt{53}$ is between 7 and 8, the largest integer $< x$ is $a_1 = 1$. Then

$$x = \frac{25 + \sqrt{53}}{22} = a_1 + \frac{1}{x_2} = 1 + \frac{1}{x_2},$$

where

$$x_2 = \frac{1}{x - 1} = \frac{22}{3 + \sqrt{53}} \, \frac{3 - \sqrt{53}}{3 - \sqrt{53}} = \frac{\sqrt{53} - 3}{2} > 1.$$

The largest integer $< x_2$ is $a_2 = 2$, so

$$x_2 = a_2 + \frac{1}{x_3} = 2 + \frac{1}{x_3},$$

where

$$x_3 = \frac{1}{x_2 - 2} = \frac{2}{\sqrt{53} - 7} = \frac{\sqrt{53} + 7}{2}.$$

The largest integer $< x_3$ is $a_3 = 7$, so

$$x_3 = a_3 + \frac{1}{x_4} = 7 + \frac{1}{x_4},$$

where

$$x_4 = \frac{1}{x_3 - 7} = \frac{2}{\sqrt{53} - 7} = \frac{\sqrt{53} + 7}{2}.$$

Thus $x_4 = x_3$, and so the last calculation will repeat over and over again.

Hence, the required expansion is

$$x = 1 + \frac{1}{x_2} = 1 + \cfrac{1}{2 + \cfrac{1}{x_3}} = 1 + \cfrac{1}{2 + \cfrac{1}{7 + \cfrac{1}{x_4}}} = \cdots,$$

so that finally we obtain

$$x = \frac{25 + \sqrt{53}}{22} = 1 + \frac{1}{2+}\,\frac{1}{7+}\,\frac{1}{7+}\cdots = [1, 2, 7, 7, \cdots]$$

$$= [1, 2, \overline{7}].$$

Now let us reverse the process; let us start with the infinite expansion and try to get back to the original value of x. It is convenient to replace

$$x = 1 + \cfrac{1}{2 + \cfrac{1}{7 + \cfrac{1}{7 + \cfrac{1}{7 + \ddots}}}} \quad \text{by} \quad x = 1 + \cfrac{1}{2 + \cfrac{1}{y}},$$

where

$$y = 7 + \cfrac{1}{7 + \cfrac{1}{7 + \ddots}} = 7 + \frac{1}{y}.$$

Then y satisfies the equation

$$y^2 - 7y - 1 = 0.$$

Solving for y (by the quadratic formula) and noting that $y > 0$, we find that

$$y = \frac{7 + \sqrt{53}}{2}.$$

Hence

$$x = 1 + \cfrac{1}{2 + \cfrac{1}{y}} = 1 + \cfrac{1}{2 + \cfrac{2}{7 + \sqrt{53}}}.$$

Simplifying the right-hand side, we obtain

$$x = \frac{23 + 3\sqrt{53}}{16 + 2\sqrt{53}} \cdot \frac{16 - 2\sqrt{53}}{16 - 2\sqrt{53}} = \frac{25 + \sqrt{53}}{22},$$

which is the original value of x.

3.3 Convergents

The convergents to the infinite continued fraction

$$x = a_1 + \frac{1}{a_2 +} \frac{1}{a_3 +} \cdots = [a_1, a_2, a_3, \cdots]$$

are calculated in exactly the same way as before. The convergent $c_n = p_n/q_n$ is calculated by the same formulas

$$p_n = a_n p_{n-1} + p_{n-2},$$
$$q_n = a_n q_{n-1} + q_{n-2}$$

for all $n \geq 1$, where, as before, we define $p_{-1} = 0$, $p_0 = 1$, $q_{-1} = 1$, and $q_0 = 0$. The computational scheme is the same.

EXAMPLE 1. The infinite continued fraction for $\pi = 3.14159 \cdots$ starts out as follows:

$$\pi = [3, 7, 15, 1, 292, 1, 1, \cdots].$$

Calculate the first five convergents. These convergents give successively better approximations to π.

SOLUTION. The table of convergents is as follows:

i	-1	0	1	2	3	4	5
a_i			3	7	15	1	292
p_i	0	1	3	22	333	355	103993
q_i	1	0	1	7	106	113	33102
c_i			$\frac{3}{1}$	$\frac{22}{7}$	$\frac{333}{106}$	$\frac{355}{113}$	$\frac{103993}{33102}$

In this connection it is interesting to note that the earliest approximation to π is to be found in the *Rhind Papyrus* preserved in the British Museum and dated about 1700 B.C. Translated into our decimal notation, the value of π stated there is 3.1604.

The approximation $\pi = 3$, less accurate than the above Egyptian value, was used by the Babylonians. Archimedes (c. 225 B.C.) stated that the ratio of the circumference of any circle to its diameter is less than $3\frac{1}{7} = \frac{22}{7} = 3.14285 \cdots$, but greater than $3\frac{10}{71} = 3.14084 \cdots$; this is quite a remarkable result considering the very limited means at his disposal. The approximation

$$\frac{355}{113} = 3.141592 \cdots$$

is correct to six decimal places. More information on the use of continued fractions to give rational approximations to irrational numbers will be taken up in Chapter 5.

Problem Set 9

1. Verify the following expansions and calculate the first five convergents:

(a) $\sqrt{6} = [2, 2, 4, 2, 4, \cdots] = [2, \overline{2, 4}]$ (d) $\dfrac{24 - \sqrt{15}}{17} = [1, 5, \overline{2, 3}]$

(b) $\sqrt{7} = [2, \overline{1, 1, 1, 4}]$ (e) $\dfrac{\sqrt{30} - 2}{13} = [0, 3, \overline{1, 2, 1, 4}]$

(c) $\sqrt{43} = [6, \overline{1, 1, 3, 1, 5, 1, 3, 1, 1, 12}]$

2. As in the second half of Example 2, Section 3.2, verify that the following continued fractions represent the irrational numbers written on the right.

(a) $[2, \overline{2, 4}] = \sqrt{6}$ (b) $[5, \overline{1, 1, 1, 10}] = \sqrt{32}$

3. *Discussion Problem.* The following is one of the classical straight-edge and compass problems. Construct, *using only a straightedge and compass*, a square equal in area to a circle of radius 1. A circle of radius 1 has an area $A = \pi r^2 = \pi$, so a square with the same area would have a side equal to $\sqrt{\pi}$. Were it possible to construct the length π we could then construct $\sqrt{\pi}$ by the following means: Let $AB = \pi$, $BC = 1$ and draw a semicircle with center at O and passing through A and C; see Figure 2. Draw BD perpendicular to AC. Then $x = BD = \sqrt{\pi}$. To prove this use the similar triangles ABD and CBD.

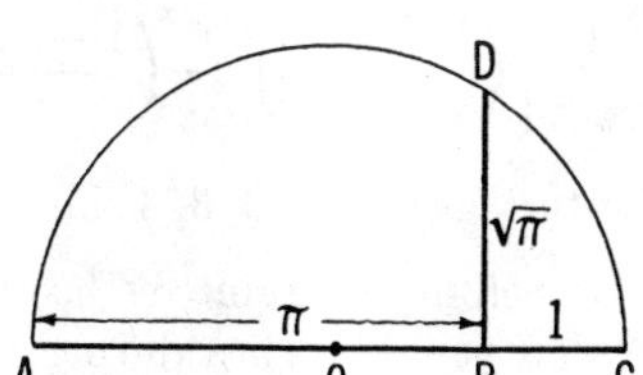

Figure 2

It can be proved that a length equal to π cannot be constructed with straightedge and compass. However, there are many interesting *approximate constructions.* For example, Jakob de Gelder in 1849 gave the following construction using the convergent $\frac{355}{113} = 3.141592 \cdots$ discussed at the end of Section 3.3. Since

$$\frac{355}{113} = 3 + \frac{4^2}{7^2 + 8^2},$$

the approximation to π can easily be constructed as follows: Let O be the center of a circle with radius $OE = 1$. Let AB be a diameter perpendicular to OE. Let $OD = \frac{7}{8}$, and $AF = \frac{1}{2}$; see Figure 3. Draw FG parallel to EO and FH parallel to DG. Then prove that $AH = 4^2/(7^2 + 8^2)$, and it only remains to construct a line equal in length to $3 + AH$.

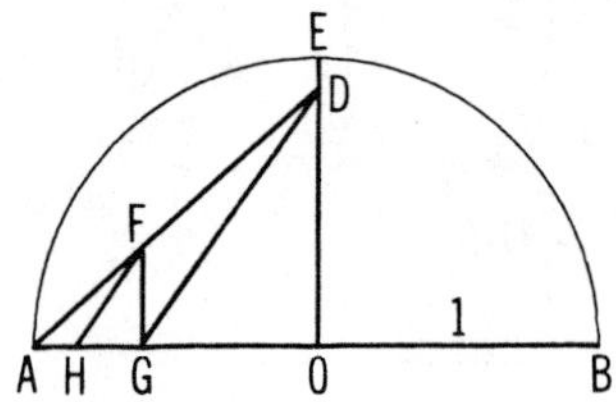

Figure 3

4. Show that $\frac{1}{2}(\sqrt{5} + 1) = [1, 1, 1, 1, \cdots]$, and also verify that the convergents are

$$\frac{1}{1}, \frac{2}{1}, \frac{3}{2}, \frac{5}{3}, \frac{8}{5}, \frac{13}{8}, \cdots,$$

both numerators and denominators being formed from the sequence of *Fibonacci numbers*

$$1, 1, 2, 3, 5, 8, 13, 21, 34, 55, \cdots.$$

Each of these numbers is the sum of the preceding two. A discussion of these interesting numbers will be given in Section 3.10.

5. The Fibonacci numbers $F_1 = 1$, $F_2 = 1$, $F_3 = 2$, $F_4 = 3$, $\cdots$ can be reproduced by substituting $n = 1, 2, 3, \cdots$ in the general formula

$$F_n = \frac{1}{\sqrt{5}}\left[\left(\frac{1 + \sqrt{5}}{2}\right)^n - \left(\frac{1 - \sqrt{5}}{2}\right)^n\right].$$

Verify this by substituting $n = 1, 2, 3, 4$ into this formula.

6. Imagine that each branch of a certain tree has the following pattern of growth. It produces no new branches during its first year of growth. During the second, it puts forth one branch, then "rests" for a year, then branches again, and so on. Sketch such a tree after a five-year growing

period and show that, if we regard the trunk and its extensions as branches, then in the first year of the tree's growth it has one branch (the trunk), in the second year two branches, and in general the number of branches will reproduce the Fibonacci numbers 1, 2, 3, 5, 8, $\cdots$.

7. *Wythoff's game* (invented in 1907 by W. A. Wythoff). Alternately two players A and B remove counters from two heaps according to the following rules: At his turn a player may take any number of counters from the first or from the second heap. If he wishes to take counters from both heaps, then he must remove an equal number of counters from each. The player who takes the last counter from the table wins.

   In order for player A to win he should, after his move, leave one of the following safe combinations (safe for A):

   $$(1, 2),\ (3, 5),\ (4, 7),\ (6, 10),\ (8, 13),\ (9, 15),\ (11, 18),\ (12, 20),\ \cdots .$$

   Then no matter what B does in the next move he will leave an unsafe combination (unsafe for B), and A can always convert this back into a safe combination (safe for A). So unless A makes a mistake, he will win the game.

   It can be proved that the nth pair of numbers forming a safe combination is given by

   $$(\{n\tau\},\ \{n\tau^2\}), \qquad n = 1, 2, 3, \cdots ,$$

   where $\tau = \frac{1}{2}(\sqrt{5} + 1)$ and where $\{x\}$ stands for the greatest integer less than or equal to x. Verify this statement for $n = 1, 2, 3$. For more details about this game and related subjects see H. S. M. Coxeter: *The Golden Section, Phyllotaxis, and Wythoff's Game*, Scripta Mathematica, vol. 19 (1953), pp. 135–143.

8. Using only a straightedge and compass, construct a point G on a line segment AB such that $(AG) = \tau(GB)$, where $\tau = \frac{1}{2}(1 + \sqrt{5})$.

9. Use the results of Problem 8 to show how to construct a regular pentagon using only a straightedge and compass.

3.4 Additional Theorems on Convergents

The numerators p_n and denominators q_n of the convergents $c_n = p_n/q_n$ of the infinite simple continued fraction

$$[a_1, a_2, \cdots, a_n, \cdots]$$

satisfy the fundamental recurrence relation

$$p_n q_{n-1} - p_{n-1} q_n = (-1)^n, \qquad n \geq 0, \tag{3.2}$$

proved in Theorem 1.4, the proof given there being independent of whether the continued fraction was finite or infinite.

From this equation, upon dividing both sides by $q_n q_{n-1}$, we find that

$$\frac{p_n}{q_n} - \frac{p_{n-1}}{q_{n-1}} = \frac{(-1)^n}{q_n q_{n-1}}, \qquad n \geq 2. \tag{3.3}$$

Since $c_n = p_n/q_n$, equation (3.3) can be stated as

THEOREM 3.1. $$c_n - c_{n-1} = \frac{(-1)^n}{q_n q_{n-1}}, \qquad n \geq 2.$$

Similarly we can prove

THEOREM 3.2. $$c_n - c_{n-2} = \frac{a_n(-1)^{n-1}}{q_n q_{n-2}}, \qquad n \geq 3.$$

PROOF. Clearly

$$c_n - c_{n-2} = \frac{p_n}{q_n} - \frac{p_{n-2}}{q_{n-2}} = \frac{p_n q_{n-2} - p_{n-2} q_n}{q_n q_{n-2}}.$$

In the numerator on the right, substitute

$$p_n = a_n p_{n-1} + p_{n-2}, \qquad q_n = a_n q_{n-1} + q_{n-2},$$

obtaining

$$\begin{aligned} p_n q_{n-2} - p_{n-2} q_n &= (a_n p_{n-1} + p_{n-2}) q_{n-2} - p_{n-2}(a_n q_{n-1} + q_{n-2}) \\ &= a_n(p_{n-1} q_{n-2} - p_{n-2} q_{n-1}) \\ &= a_n(-1)^{n-1}, \end{aligned}$$

where the last equality follows from equation (3.2) with n replaced by $n - 1$. This proves Theorem 3.2.

These theorems give us important information as to how the convergents c_n change as n increases. If we set $n = 2$ and then $n = 3$ in Theorem 3.1, and recall that the q_n's are positive, we see that

$$c_2 - c_1 = \frac{1}{q_2 q_1} > 0, \qquad c_3 - c_2 = \frac{-1}{q_3 q_2} < 0,$$

respectively. These inequalities show that

$$c_1 < c_2 \quad \text{and that} \quad c_3 < c_2. \tag{3.4}$$

On the other hand, setting $n = 3$ in Theorem 3.2 shows that

$$c_3 - c_1 = \frac{a_3(-1)^2}{q_3 q_1} = \frac{a_3}{q_3 q_1} > 0,$$

since q_3, q_1, a_3 are all positive numbers. Hence $c_1 < c_3$, and combining this result with those in (3.4) proves that

$$c_1 < c_3 < c_2.$$

Similarly, using $n = 3$, then $n = 4$ in Theorem 3.1, followed by $n = 4$ in Theorem 3.2, we see that

$$c_3 < c_4 < c_2.$$

Proceeding step by step in this fashion, we obtain the inequalities

$$c_3 < c_5 < c_4,$$

$$c_5 < c_6 < c_4,$$

$$\cdots\cdots\cdots$$

Combining these inequalities we obtain the fundamental result

$$c_1 < c_3 < c_5 < \cdots < c_{2n+1} < \cdots \quad \cdots < c_{2n} < \cdots < c_6 < c_4 < c_2.$$

We state it as a theorem:

THEOREM 3.3. *The odd convergents c_{2n+1} of an infinite simple continued fraction form an increasing sequence, and the even convergents c_{2n} form a decreasing sequence, and every odd convergent is less than any even convergent. Moreover, each convergent c_n, $n \geq 3$, lies between the two preceding convergents.*

Problem Set 10

1. Give a numerical verification of Theorem 3.3 using the convergents to $\sqrt{2}$.

3.5 Some Notions of a Limit

The conversion of an irrational number x into an infinite continued fraction gave, as we have seen, in succession,

$$x = a_1 + \frac{1}{x_2},$$

$$x_2 = a_2 + \frac{1}{x_3},$$

$$\cdots\cdots\cdots\cdots$$

$$x_{n-1} = a_{n-1} + \frac{1}{x_n},$$

$$\cdots\cdots\cdots\cdots\cdots,$$

so that, at the end of the $(n-1)$st calculation we had

$$x = a_1 + \frac{1}{a_2 +} \frac{1}{a_3 +} \cdots \frac{1}{+ a_{n-1} +} \frac{1}{x_n}, \tag{3.5}$$

where x_n is irrational, and we saw that the process could be continued indefinitely. Realizing this, one is tempted to write (as we did)

$$x = a_1 + \frac{1}{a_2 +} \frac{1}{a_3 +} \cdots \frac{1}{+ a_n +} \cdots,$$

which implies that the *infinite* continued fraction on the right actually represents the irrational number x. It is advisable to reflect on the *meaning* of such a statement. The implication is that we can somehow carry out an *infinite* number of operations and thereby arrive at a certain number which is asserted to be x, the given irrational number. We shall see, however, that *the only way to attach a mathematical meaning to such an infinite process is to introduce the notion of a limit.*

To make this clear, let us first go back to ordinary addition. Which of the following infinite sums have meaning?

$$A = 1 + 1 + 1 + 1 + \cdots,$$

$$B = 1 + \left(\frac{1}{2}\right)^1 + \left(\frac{1}{2}\right)^2 + \left(\frac{1}{2}\right)^3 + \cdots + \left(\frac{1}{2}\right)^n + \cdots.$$

Clearly, if we add 1 to itself over and over, we can make the "sum" as large as we please, so we say the sum A becomes infinite as the number of terms added increases indefinitely, and such a result is not of much use to us. On the other hand, if we add the numbers 1, $\frac{1}{2}$, $\frac{1}{4}$, $\frac{1}{8}$, $\cdots$ we get in succession the *partial sums*

$$s_1 = 1,$$

$$s_2 = 1 + \left(\frac{1}{2}\right)^1,$$

$$s_3 = 1 + \left(\frac{1}{2}\right)^1 + \left(\frac{1}{2}\right)^2,$$

$$\cdots\cdots\cdots\cdots\cdots\cdots\cdots$$

$$s_n = 1 + \left(\frac{1}{2}\right)^1 + \left(\frac{1}{2}\right)^2 + \left(\frac{1}{2}\right)^3 + \cdots + \left(\frac{1}{2}\right)^{n-1},$$

$$\cdots\cdots\cdots\cdots\cdots\cdots\cdots\cdots\cdots\cdots\cdots\cdots\cdots\cdots$$

which can be represented graphically as shown in Figure 4, where

$$s_1 < s_2 < s_3 < \cdots < s_n < \cdots,$$

so that the partial sums *continually increase.* But each partial sum s_n is less than 2; that is, they are all *bounded above* by the constant 2.

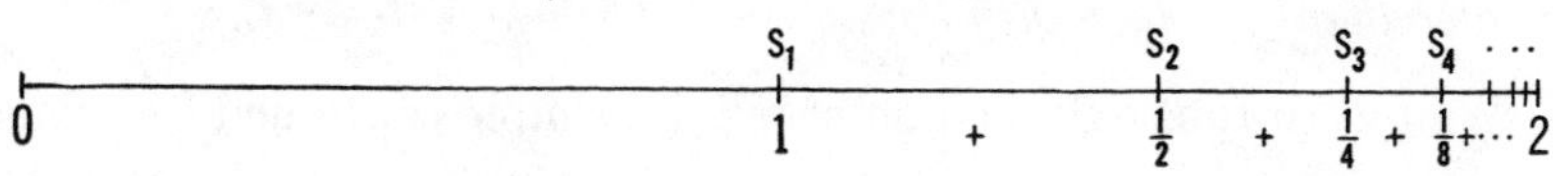

In order to prove that they continually approach this upper limit 2, we write

$$s_n = 1 + \left(\frac{1}{2}\right)^1 + \left(\frac{1}{2}\right)^2 + \cdots + \left(\frac{1}{2}\right)^{n-1},$$

so that

$$\frac{1}{2}s_n = \left(\frac{1}{2}\right) + \left(\frac{1}{2}\right)^2 + \left(\frac{1}{2}\right)^3 + \cdots + \left(\frac{1}{2}\right)^{n-1} + \left(\frac{1}{2}\right)^n.$$

Subtracting the second line from the first, we obtain

$$s_n\left(1 - \frac{1}{2}\right) = 1 - \left(\frac{1}{2}\right)^n,$$

which implies that

$$s_n = \frac{1 - (\frac{1}{2})^n}{\frac{1}{2}} = 2 - (\tfrac{1}{2})^{n-1}.$$

As n increases indefinitely, that is, as $n \to \infty$, $(\frac{1}{2})^{n-1}$ approaches zero, and so s_n gets closer and closer to 2, or approaches 2 as a *limit.*

We say that s_n *converges* to the value 2 as $n \to \infty$, or in symbols,

$$\lim_{n\to\infty} s_n = 2.$$

We then *assign* this limit 2 as the *value* of the infinite sum in question and we write

$$1 + \left(\frac{1}{2}\right) + \left(\frac{1}{2}\right)^2 + \cdots + \left(\frac{1}{2}\right)^n + \cdots = 2.$$

This illustrates, in an admittedly rough fashion, the mathematical notion of a limit needed to attach meaning to an infinite continued fraction. It also illustrates a fundamental theorem of analysis which we state but do not attempt to prove.†

THEOREM 3.4. *If a sequence of numbers* $s_1, s_2, s_3, \cdots$ *continually increases, and if for each* n, s_n *is less than* U, *where* U *is some fixed number, then the numbers* $s_1, s_2, s_3, \cdots$ *have a limit* l_U, *where* $l_U \le U$. *If the numbers* $s_1, s_2, s_3, \cdots$ *continually decrease but are all greater than* L, *then they have a limit* l_L, *where* $l_L \ge L$.

We return to the discussion of infinite simple continued fractions.

3.6 Infinite Continued Fractions

Our task is to attach a meaning to the infinite continued fraction

$$a_1 + \frac{1}{a_2 +}\,\frac{1}{a_3 +}\,\cdots\,\frac{1}{a_n +}\,\cdots.$$

Theorem 3.3 states that the odd convergents $c_1, c_3, c_5, \cdots$ form an increasing sequence of numbers all bounded above by the convergent $c_2 = U$, that is,

$$c_1 < c_3 < \cdots < c_{2n+1} < \cdots\cdots < c_{2n} < \cdots < c_4 < c_2 = U;$$

hence they will converge to a limit $l_U \le U$. Moreover, since all odd convergents are less than all the even convergents, the limit l_U must be a number less than all the even convergents.

On the other hand, the even convergents $c_2, c_4, c_6, \cdots, c_{2n}, \cdots$ form a decreasing sequence of numbers all bounded below by the convergent $c_1 = L$, that is,

† For a discussion of limits of sequences, see L. Zippin [15], which also treats this fundamental theorem of analysis (Theorem 3.4).

$$L = c_1 < c_3 < \cdots < c_{2n+1} < \cdots\cdots < c_{2n} < \cdots < c_4 < c_2,$$

so that even convergents approach the limit $l_L \geq L$, where l_L is a number greater than every odd convergent. Looking at the convergents graphically (see Figure 5), we see that *all we have proved so far* is that the even convergents have a limit l_L and the odd convergents have a limit l_U. If $l_U \neq l_L$ we would be in trouble. We can prove, however, that $l_U = l_L$.

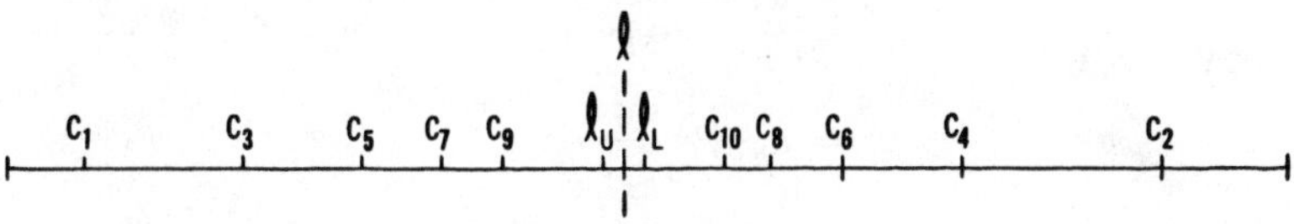

Figure 5

To this end, return to Theorem 3.1 and replace n by $2k$ and $n-1$ by $2k-1$. We get

$$c_{2k} - c_{2k-1} = \frac{(-1)^{2k}}{q_{2k}q_{2k-1}},$$

or, since $(-1)^{2k} = 1$,

$$c_{2k} - c_{2k-1} = \frac{1}{q_{2k}q_{2k-1}}. \tag{3.6}$$

The numbers q_n are calculated by means of the recurrence relation

$$q_n = a_n q_{n-1} + q_{n-2};$$

therefore it follows, since each a_n $(n \geq 2)$ and each q_n $(n \geq 1)$ is a positive integer, that the q_n's increase without bound as n increases. Hence, the denominator $q_{2k}q_{2k-1}$ of the fraction in (3.6) increases without bound as k increases, that is, the fraction $1/q_{2k}q_{2k-1}$ approaches zero as k approaches infinity. But then from equation (3.6) we conclude that the difference $c_{2k} - c_{2k-1}$ approaches zero as k approaches infinity, and the only way this can happen is for both c_{2k} and c_{2k-1} to have the same limiting value $l = l_U = l_L$. We have proved:

THEOREM 3.5. *Every infinite simple continued fraction converges to a limit l which is greater than any odd convergent and less than any even convergent.*

How far have we progressed? Is this limit l the same number x which gave rise to the continued fraction in the first place? Actually it is, but this must be proved.

To do so, let x be the given irrational number, and return to the expansion (3.5),

$$x = a_1 + \frac{1}{a_2 +}\,\frac{1}{a_3 +}\cdots\frac{1}{+ a_{n-1} +}\,\frac{1}{x_n},$$

where x_n is the "rest" of the fraction, that is,

$$\begin{aligned} x_n &= a_n + \frac{1}{a_{n+1} +}\,\frac{1}{a_{n+2} +}\cdots \\ &= a_n + \frac{1}{x_{n+1}}, \end{aligned} \tag{3.7}$$

where again

$$x_{n+1} = a_{n+1} + \frac{1}{a_{n+2} +}\cdots. \tag{3.8}$$

The second line in (3.7) shows that

$$x_n > a_n,$$

since x_{n+1} is positive. Similarly, (3.8) shows that

$$x_{n+1} > a_{n+1}, \qquad \text{or} \qquad \frac{1}{x_{n+1}} < \frac{1}{a_{n+1}}.$$

Again, according to the second line in (3.7),

$$x_n = a_n + \frac{1}{x_{n+1}},$$

and since $\frac{1}{x_{n+1}} < \frac{1}{a_{n+1}}$, it follows that

$$x_n < a_n + \frac{1}{a_{n+1}};$$

so, combining these results, we see that

$$a_n < x_n < a_n + \frac{1}{a_{n+1}}. \tag{3.9}$$

The next step in the proof is to show that x lies between c_n and c_{n+1}. To this end, we compare the three expressions:

$$c_n = a_1 + \frac{1}{a_2} + \cdots + \frac{1}{a_{n-1}} + \frac{1}{a_n},$$

$$(3.10) \qquad x = a_1 + \frac{1}{a_2} + \cdots + \frac{1}{a_{n-1}} + \frac{1}{x_n},$$

$$c_{n+1} = a_1 + \frac{1}{a_2} + \cdots + \frac{1}{a_{n-1}} + \frac{1}{a_n} + \frac{1}{a_{n+1}}.$$

We first observe that these expressions have the term

$$a_1 + \frac{1}{a_2} + \cdots + \frac{1}{a_{n-1}}$$

in common so that it is necessary only to compare the terms in which they differ, namely

$$\frac{1}{a_n}, \qquad \frac{1}{x_n}, \qquad \text{and} \qquad \frac{1}{a_n + \dfrac{1}{a_{n+1}}}.$$

But by (3.9) we know that

$$\frac{1}{a_n} > \frac{1}{x_n} > \frac{1}{a_n + \dfrac{1}{a_{n+1}}},$$

and we can conclude from (3.10) that x will always lie between two consecutive convergents c_n and c_{n+1}; that is, either

$$c_n < x < c_{n+1} \qquad \text{or} \qquad c_n > x > c_{n+1}.$$

A direct calculation shows that

$$c_1 < x < c_2;$$

for, (3.9) gives $a_1 < x_1$, and since $c_1 = a_1$ and $x_1 = x$, we see that $c_1 < x$. On the other hand, $x = a_1 + 1/x_2$, where by (3.9) $a_2 < x_2$ or $1/x_2 < 1/a_2$; hence

$$x = a_1 + \frac{1}{x_2} < a_1 + \frac{1}{a_2} = c_2.$$

Thus

$$c_1 < x < c_2.$$

Similarly, equations (3.10) show that x lies between c_2 and c_3, between c_3 and c_4, between c_4 and c_5, and so on. Since all odd convergents are less than all even convergents, we are forced to the conclusion that

$$c_{2k-1} < x < c_{2k}, \qquad k = 1, 2, 3, \cdots,$$

or, in expanded form, that

$$c_1 < c_3 < \cdots < c_{2k-1} < \cdots < x < \cdots < c_{2k} < \cdots < c_4 < c_2.$$

Thus we see that the convergents $c_1, c_3, \cdots$ approach x from the left, and $c_2, c_4, \cdots$ approach x from the right. But we know that as k increases indefinitely, the odd convergents c_{2k-1} and the even convergents c_{2k} approach a limit l; hence x and l must be one and the same. Therefore it is permissible to write

$$x = a_1 + \frac{1}{a_2 +} \cdots \frac{1}{+ a_n +} \cdots,$$

and we have proved

THEOREM 3.6. *If an irrational number x is expanded into an infinite simple continued fraction $[a_1, a_2, \cdots, a_n, \cdots]$ according to the rules described, then the limit to which the convergents $c_1, c_2, \cdots, c_n, \cdots$ of the fraction $[a_1, a_2, \cdots, a_n, \cdots]$ converge is the number x which gave rise to the fraction in the first place.*

This theorem should be followed by an additional theorem stating that the expansion of any irrational number into an infinite simple continued fraction is unique. This is true, and the reader will find it impossible to expand, as explained, any given irrational numbers in *two* different ways.

3.7 Approximation Theorems

Our experience with continued fractions and in particular our study of Theorem 3.6 have supplied ample evidence that each convergent in the continued fraction expansion of an irrational number x is nearer to the value of x than is the preceding convergent. Before stating such a result as a theorem we make some preliminary remarks.

Let the expansion of the irrational number x be

$$x = a_1 + \frac{1}{a_2 +} \frac{1}{a_3 +} \cdots \frac{1}{+ a_n +} \frac{1}{x_{n+1}}, \tag{3.11}$$

where

$$x_{n+1} = a_{n+1} + \frac{1}{a_{n+2}} + \frac{1}{a_{n+3}} + \cdots .$$

We assume that $x_2, x_3, \cdots$ are all positive numbers; also note that $x_1 = x$. While x_{n+1} contains an infinite number of integral partial quotients $a_{n+1}, a_{n+2}, \cdots$, it need not itself be an integer, and consequently we have no right to treat it as though it were a legitimate partial quotient.

Suppose, however, we write (3.11) in the form

$$x = [a_1, a_2, \cdots, a_n, x_{n+1}]$$

of a "finite" continued fraction and treat x_{n+1} as a legitimate partial quotient. Then, if we calculate convergents in the usual manner, the last "convergent" (in Theorem 1.3, take $i = n + 1$ and $a_{n+1} = x_{n+1}$) would be

$$\frac{x_{n+1}p_n + p_{n-1}}{x_{n+1}q_n + q_{n-1}},$$

and, by analogy with our study of finite continued fractions, this should be equal to x, the given irrational number. Thus it seems reasonable to write

$$x = [a_1, a_2, \cdots, a_n, x_{n+1}] = \frac{x_{n+1}p_n + p_{n-1}}{x_{n+1}q_n + q_{n-1}}, \tag{3.12}$$

where, it should be stressed, p_n, q_n, p_{n-1}, q_{n-1} depend only upon the integers $a_1, a_2, \ldots, a_n$ as before. In particular, when $n = 0$, equation (3.12) gives

$$\frac{x_1p_0 + p_{-1}}{x_1q_0 + q_{-1}} = \frac{x_1 \cdot 1 + 0}{x_1 \cdot 0 + 1} = x_1$$

and by definition,

$$x_1 = a_1 + \frac{1}{a_2} + \cdots = x.$$

When $n = 1$, (3.12) gives

$$[a_1, x_2] = \frac{x_2p_1 + p_0}{x_2q_1 + q_0} = \frac{x_2 \cdot a_1 + 1}{x_2 \cdot 1 + 0} = a_1 + \frac{1}{x_2}$$

$$= a_1 + \frac{1}{a_2} + \cdots = x.$$

That (3.12) holds for all n can be proved in exactly the same way as we proved Theorem 1.3, the successive steps being nearly identical. We are now ready to state the main theorem of this section:

THEOREM 3.7. *Each convergent is nearer to the value of an infinite simple continued fraction than is the preceding convergent.*

PROOF. Let the expansion of the given irrational number x be

$$x = [a_1, a_2, \cdots, a_n, x_{n+1}],$$

where

$$x_{n+1} = [a_{n+1}, a_{n+2}, \cdots].$$

Then, according to (3.12),

$$x = \frac{x_{n+1}p_n + p_{n-1}}{x_{n+1}q_n + q_{n-1}},$$

and from this we obtain

$$x(x_{n+1}q_n + q_{n-1}) = x_{n+1}p_n + p_{n-1},$$

or, rearranging, we have for $n \geq 2$

$$\begin{aligned} x_{n+1}(xq_n - p_n) &= -(xq_{n-1} - p_{n-1}) \\ &= -q_{n-1}\left(x - \frac{p_{n-1}}{q_{n-1}}\right). \end{aligned}$$

Dividing through by $x_{n+1}q_n$, we obtain

$$x - \frac{p_n}{q_n} = \left(-\frac{q_{n-1}}{x_{n+1}q_n}\right)\left(x - \frac{p_{n-1}}{q_{n-1}}\right).$$

Now if $a = b \cdot c$, then $|a| = |b| \cdot |c|$, and $|-a| = |a|$;† hence

$$\left|x - \frac{p_n}{q_n}\right| = \left|\frac{q_{n-1}}{x_{n+1}q_n}\right| \cdot \left|x - \frac{p_{n-1}}{q_{n-1}}\right|. \tag{3.13}$$

We know that for $n \geq 2$, $x_{n+1} > 1$, and that $q_n > q_{n-1} > 0$; hence

$$0 < \frac{q_{n-1}}{x_{n+1}q_n} < 1,$$

† The symbol $|a|$, read "absolute value of a", means

$$\begin{aligned} |a| &= a && \text{if } a \geq 0; \\ |a| &= -a && \text{if } a < 0. \end{aligned}$$

For example, $|7| = 7$, $|-7| = 7$.

and so

$$0 < \left|\frac{q_{n-1}}{x_{n+1}q_n}\right| < 1.$$

Thus (3.13) shows that

$$\left|x - \frac{p_n}{q_n}\right| < \left|x - \frac{p_{n-1}}{q_{n-1}}\right|, \qquad n \geq 2,$$

or, what is the same thing,

$$|x - c_n| < |x - c_{n-1}|, \qquad n \geq 2,$$

This shows that c_n is closer to x than is c_{n-1}, and the theorem is proved.

It would be interesting to have some *measure*, or *estimate*, of just how closely c_n approximates x. In fact, we know already from Theorem 3.1, with n replaced by $n + 1$, that

$$c_{n+1} - c_n = \frac{(-1)^{n+1}}{q_{n+1}q_n}.$$

Taking the absolute value of both sides, this tells us that

$$|c_{n+1} - c_n| = \frac{1}{q_{n+1}q_n}, \qquad n \geq 1.$$

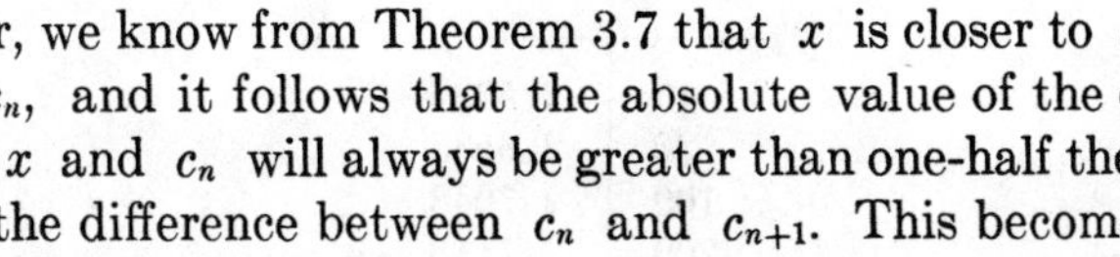

Figure 6

Moreover, we know from Theorem 3.7 that x is closer to c_{n+1} than it is to c_n, and it follows that the absolute value of the difference between x and c_n will always be greater than one-half the absolute value of the difference between c_n and c_{n+1}. This becomes clear if the situation is studied graphically. Figure 6 shows the case *when n is odd,* so that c_n is to the left of c_{n+1}. Clearly $AB < AC < AD$, or

$$\frac{1}{2q_nq_{n+1}} < \left|x - \frac{p_n}{q_n}\right| < \frac{1}{q_nq_{n+1}}.$$

Since $q_{n+1} > q_n$, $q_nq_{n+1} > q_n^2$ and so $1/q_nq_{n+1} < 1/q_n^2$. Hence we can state

THEOREM 3.8. $$\frac{1}{2q_nq_{n+1}} < \left|x - \frac{p_n}{q_n}\right| < \frac{1}{q_nq_{n+1}} < \frac{1}{q_n^2}, \qquad n \geq 1.$$

If x is irrational, there exists an infinite number of convergents p_n/q_n satisfying Theorem 3.8. Thus we have the following theorem:

THEOREM 3.9. *If x is irrational, there exists an infinite number of rational fractions p/q, $q > 0$, $(p, q) = 1$, such that*

$$\left|x - \frac{p}{q}\right| < \frac{1}{q^2}.$$

This is the beginning of the theory of rational approximation to irrational numbers, a subject we shall discuss briefly in Chapter 5.

EXAMPLE 1. Show that the first few convergents to the number

$$e = 2.718282 \cdots$$

give better and better approximations to this number. These convergents should be calculated by finding the first few convergents to 2.718282, a decimal fraction which approximates e correctly to six decimal places.

Comments. The irrational number e arises quite naturally in the study of calculus and is defined as

$$e = \lim_{n\to\infty}\left(1 + \frac{1}{n}\right)^n.$$

That the sequence of numbers $\left(1 + \frac{1}{1}\right)^1, \left(1 + \frac{1}{2}\right)^2, \ldots, \left(1 + \frac{1}{n}\right)^n, \ldots$

actually approaches a limit can be suggested by numerical evidence:

n	$\left(1 + \frac{1}{n}\right)^n$
10	$2.5937 \cdots$
20	$2.6533 \cdots$
100	$2.7048 \cdots$
200	$2.7115 \cdots$
1000	$2.7169 \cdots$
....	....

The number e is taken as the base of the system of *natural logarithms*, just as 10 is used as the base for *common logarithms*. The continued fraction expansion of e is

$$e = [2, 1, 2, 1, 1, 4, 1, 1, 6, 1, 1, 8, \cdots];$$

the proof is quite difficult.

SOLUTION. Assuming the above expansion for e, or being content with the approximation

$$e = 2.718282 = \frac{1359141}{500000},$$

we find that

$$e = [2, 1, 2, 1, 1, 4, 1, \cdots].$$

The corresponding convergents are

$$\frac{2}{1}, \frac{3}{1}, \frac{8}{3}, \frac{11}{4}, \frac{19}{7}, \frac{87}{32}, \frac{106}{39}, \dots,$$

and a conversion to decimals shows, indeed, that in succession these give better and better approximations to e.

As a check on Theorem 3.9, notice that $p_7/q_7 = \frac{106}{39}$; hence it should be true that

$$\left|e - \frac{p_7}{q_7}\right| < \frac{1}{q_7^2},$$

or

$$\left|e - \frac{106}{39}\right| < \frac{1}{39^2}.$$

A numerical calculation shows that

$$e - \frac{106}{39} = 0.00033264 \cdots,$$

and this is certainly less than $1/39^2 = 0.00065746\cdots$. We observe that the value of $e - \frac{106}{39}$ is approximately one-half that of $1/39^2$, and this suggests that Theorem 3.9, regarded as an approximation theorem, might be considerably improved. We shall see in Chapter 5 that this is indeed the case.

The inequality

$$\left|x - \frac{p_n}{q_n}\right| < \frac{1}{q_n^2}$$

of Theorem 3.8 is true for rational or irrational x. In the next example we shall approximate a rational number.

EXAMPLE 2. Given the fraction $\frac{2065}{902}$, find a fraction with a smaller numerator and a smaller denominator whose value approximates that of the given fraction correctly to three decimal places.

SOLUTION. Convert $\frac{2065}{902}$ into a continued fraction and calculate the convergents. The table gives the numerical results:

i	-1	0	1	2	3	4	5	6	7
a_i			2	3	2	5	5	1	3
p_i	0	1	2	7	16	87	451	538	2065
q_i	1	0	1	3	7	38	197	235	902
c_i			$\frac{2}{1}$	$\frac{7}{3}$	$\frac{16}{7}$	$\frac{87}{38}$	$\frac{451}{197}$	$\frac{538}{235}$	$\frac{2065}{902}$

Referring now to Theorem 3.8, we search for two convergents $c_n = p_n/q_n$ and $c_{n+1} = p_{n+1}/q_{n+1}$ which will make

$$\left|\frac{2065}{902} - \frac{p_n}{q_n}\right| < \frac{1}{q_n q_{n+1}} < 0.0005.$$

That is, we wish to approximate $\frac{2065}{902}$ by p_n/q_n with an *error* less than half a unit in the fourth decimal place. A little experimentation soon shows that

$$\frac{87}{38} = \frac{p_4}{q_4}, \qquad \frac{451}{197} = \frac{p_5}{q_5}$$

will suffice, for

$$\left|\frac{2065}{902} - \frac{87}{38}\right| < \frac{1}{q_4 q_5} = \frac{1}{38 \cdot 197} < 0.00013.$$

Hence the required fraction is $\frac{87}{38}$. Note that if we had worked with the fraction $1/q_n^2$ instead of $1/q_n q_{n+1}$ our answer would have been the next convergent $\frac{451}{197}$, since $1/38^2$ is not less than 0.0005. In order to find values of $q_n q_{n+1}$ such that $1/q_n q_{n+1} < \epsilon$, where ϵ is any given number, we could use a table of squares and first check that $q_n^2 > 1/\epsilon$, following this by an additional check to see if $q_n q_{n+1} > 1/\epsilon$.

Problem Set 11

1. Given the fraction $\frac{2893}{1323}$, find a fraction with a smaller numerator and a smaller denominator whose value approximates that of the given fraction correctly to three decimal places, that is, with an error of less than 5 units in the fourth place.

2. Expand $\sqrt{19}$ into an infinite simple continued fraction and find a fraction which will approximate $\sqrt{19}$ with accuracy to four decimal places.

3. The continued fraction expansion of π is $[3, 7, 15, 1, 292, 1, 1, 1, \cdots]$. Use Theorem 3.8 to investigate how closely the first four convergents approximate π.

3.8 Geometrical Interpretation of Continued Fractions

A striking geometrical interpretation of the manner in which the convergents $c_1, c_2, \cdots, c_n, \cdots$ of a continued fraction for an irrational number converge to the value of the given number was given by Felix Klein† in 1897. Felix Klein was not only a prominent mathematician but a most popular mathematical expositor, some of whose works are available today in reprint form.

Let α be an irrational number whose expansion is

$$[a_1, a_2, \cdots, a_n, \cdots],$$

and whose convergents are

$$c_1 = \frac{p_1}{q_1}, \qquad c_2 = \frac{p_2}{q_2}, \qquad \cdots, \qquad c_n = \frac{p_n}{q_n}, \qquad \cdots .$$

For simplicity, assume α positive, and on graph paper mark with dots all points (x, y) whose coordinates x and y are positive integers. At these points, called *lattice points*, imagine that pegs or pins are inserted. Next plot the line

$$y = \alpha x.$$

This line does not pass through any of the lattice points; for, if it did there would be a point (x, y) with *integral* coordinates satisfying the equation $y = \alpha x$, and $\alpha = y/x$ would be a rational number. This is impossible since α is irrational.

Now imagine that a piece of thin black thread is tied to an infinitely remote point on the line $y = \alpha x$, and that we hold the other end of the thread in our hand. We pull the thread taut so that the end in our hand is at the origin. Keeping the thread taut, we move our hand away from the origin, toward the left; the thread will catch on certain pegs above the line. If we move the thread away from the line in the other direction, it will catch on certain other pegs. See Figure 7.

The pegs contacted by the thread on the lower side are situated at the lattice points with coordinates

$$(q_1, p_1), \qquad (q_3, p_3), \qquad (q_5, p_5), \qquad \cdots$$

and correspond, respectively, to the odd convergents,

$$c_1 = \frac{p_1}{q_1}, \qquad c_3 = \frac{p_3}{q_3}, \qquad c_5 = \frac{p_5}{q_5}, \qquad \cdots,$$

† F. Klein: *Ausgewählte Kapitel der Zahlentheorie*, Teubner, 1907, pp. 17–25.

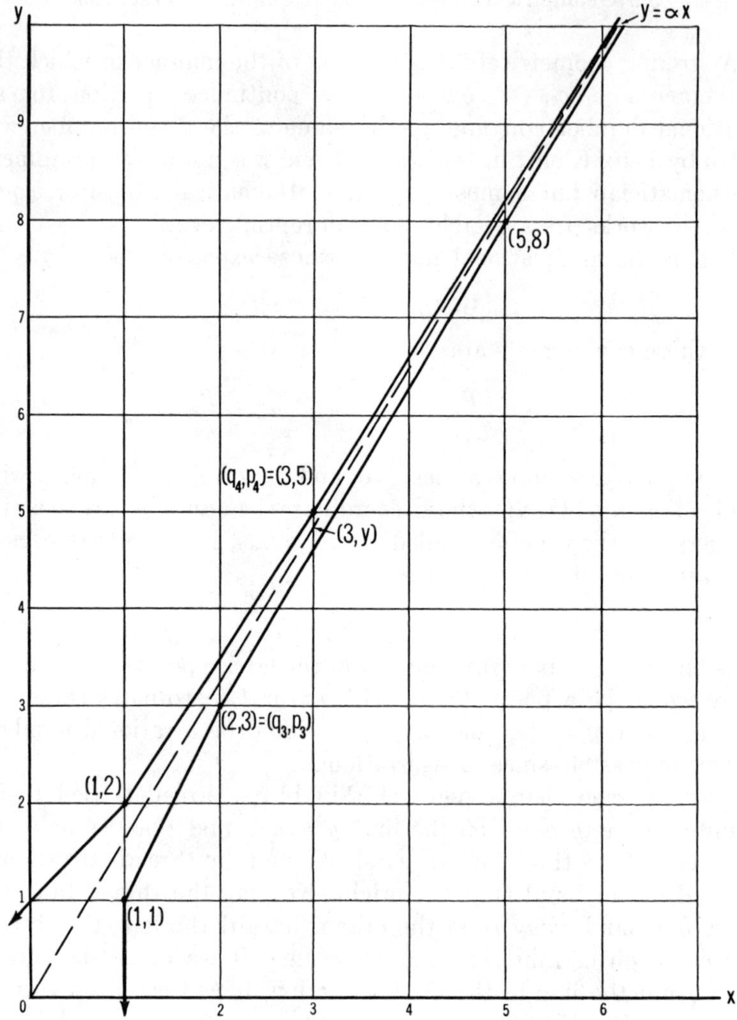

Figure 7

which are all less than α. The pegs contacted above the line are situated at the lattice points

$$(q_2, p_2), \qquad (q_4, p_4), \qquad (q_6, p_6), \qquad \cdots$$

corresponding to the even convergents,

$$c_2 = \frac{p_2}{q_2}, \qquad c_4 + \frac{p_4}{q_4}, \qquad c_6 = \frac{p_6}{q_6}, \qquad \cdots,$$

all of which are greater than α. Each of the two positions of the string forms a *polygonal* path which approaches the line $y = \alpha x$ more and more closely the farther out we go.

EXAMPLE. Draw a Klein diagram for the continued fraction expansion of

$$\alpha = \frac{1 + \sqrt{5}}{2} = [1, 1, 1, 1, \cdots].$$

SOLUTION. The convergents are

$$\frac{1}{1}, \frac{2}{1}, \frac{3}{2}, \frac{5}{3}, \frac{8}{5}, \frac{13}{8}, \cdots.$$

The points or pegs corresponding to the odd convergents are (1, 1), (2, 3), (5, 8), $\cdots$ and are all below the line; see Figure 7. Those points corresponding to the even convergents are (1, 2), (3, 5), (8, 13), $\cdots$ and are above the line.

Let us show, for example, that the point $(q_4, p_4) = (3, 5)$ corresponds to the even convergent $p_4/q_4 = 5/3$, which is greater than α. Consider the point $(3, y)$ marked in Figure 7. Since it is on the line $y = \alpha x$, we see that $y = \alpha \cdot 3$, or $\alpha = y/3$. The point (3, 5) is above the line so $5 > y$, or $5/3 > y/3 = \alpha$; hence the convergent $5/3 > \alpha$.

Most of the elementary properties of continued fractions have geometrical interpretations. In fact, the theory of simple continued fractions can be developed geometrically.†

Problem Set 12

1. Construct a Klein diagram for the continued fraction expansion of $(\sqrt{5} - 1)/2$.

2. Construct a Klein diagram for the continued fraction expansion of $\sqrt{3}$.

† See H. Hancock, *Development of the Minkowski Geometry of Numbers*, New York: The Macmillan Company, 1939, (Chapter 8).

3.9 Solution of the Equation $x^2 = ax + 1$

Continued fractions can be used to approximate the positive root of any polynomial equation, provided, of course, that it has such a root. We shall now examine the quadratic polynomial equation

$$x^2 = ax + 1. \tag{3.14}$$

If $a > 0$, the positive root of any quadratic equation of the form (3.14) has the continued fraction expansion

$$x = a + \frac{1}{a+}\,\frac{1}{a+}\cdots.$$

To see this, we have only to divide both sides of (3.14) by x, getting

$$x = a + \frac{1}{x},$$

so that

$$x = a + \cfrac{1}{a + \cfrac{1}{x}} = a + \cfrac{1}{a + \cfrac{1}{a + \ddots}}.$$

For example, when $a = 1$, the equation

$$x^2 = x + 1$$

has a positive root

$$x = [1, 1, 1, 1, \cdots],$$

and the successive convergents to this continued fraction will give better and better approximations to the actual solution $\frac{1}{2}(1 + \sqrt{5})$. See also Problem 4 of Section 3.3. A more detailed discussion of this particular number follows in the next section.

Problem Set 13

1. Use the quadratic formula to find the positive roots of the following equations and compare the exact solutions with the approximate solutions obtained by computing the first few convergents to the continued fraction expansions of these positive roots.

   (a) $x^2 - 3x - 1 = 0$ (b) $x^2 - 5x - 1 = 0$

2. Suppose that

$$x = b + \frac{1}{a+}\,\frac{1}{b+}\,\frac{1}{a+}\cdots = [\overline{b,\, a}]$$

and that b is a multiple of a, that is $b = ac$ (where c is an integer). Show that then x satisfies the equation

$$x^2 - bx - c = 0$$

and has the value

$$x = \frac{b + \sqrt{b^2 + 4c}}{2}.$$

3. Verify, by giving the positive integers a and b particular values, and by selecting particular convergents p_{n-2}/q_{n-2}, p_n/q_n, p_{n+2}/q_{n+2}, that if

$$x = \frac{1}{a} + \frac{1}{b+}\,\frac{1}{a+}\,\frac{1}{b+}\,\frac{1}{a+}\,\frac{1}{b+}\cdots,$$

then

$$p_{n+2} - (ab + 2)p_n + p_{n-2} = 0.$$

3.10 Fibonacci Numbers

The simplest of all infinite simple continued fractions is

$$\tau = [1, 1, 1, \cdots],$$

where τ satisfies the equation

$$\tau = 1 + \frac{1}{\tau}, \qquad \text{or} \qquad \tau^2 - \tau - 1 = 0,$$

which has the positive root

$$\tau = \frac{1 + \sqrt{5}}{2}.$$

The convergents to τ are

(3.15) $$\frac{1}{1}, \frac{2}{1}, \frac{3}{2}, \frac{5}{3}, \frac{8}{5}, \frac{13}{8}, \ldots,$$

both numerators and denominators being formed from the sequence of integers

(3.16) $$1,\ 1,\ 2,\ 3,\ 5,\ 8,\ 13,\ 21,\ 34,\ \cdots$$

Each of these numbers, after the first two, is equal to the sum of the preceding two; thus $2 = 1 + 1$, $3 = 2 + 1$, and so on. The numbers (3.16) are known as the *Fibonacci numbers*, named after the great thirteenth century mathematician Leonardo Fibonacci (c. 1170–1250), although he was not the first to use them.

The Greeks claimed that the creations of nature and art owed their beauty to certain underlying mathematical patterns. One of these was the law of the *golden mean*, or *golden section*, which has many forms. In geometry, it arises from what some call the "most pleasing" division of a line segment AB by a point C. This is said to be attained by selecting a point C such that the ratio of the parts a to b (see Figure 8) is the same as the ratio of b to the whole segment $a + b$, i.e.,

$$\frac{a}{b} = \frac{b}{a+b}, \qquad \text{or} \qquad \frac{b}{a} = \frac{a}{b} + 1.$$

If we now let $x = b/a$, we have

$$x = \frac{1}{x} + 1, \qquad \text{or} \qquad x^2 - x - 1 = 0,$$

so that $x = b/a = \frac{1}{2}(1 + \sqrt{5}) = \tau$, or $b = \tau a$. Thus a line segment is said to be divided according to the golden mean if one part is τ times the other.

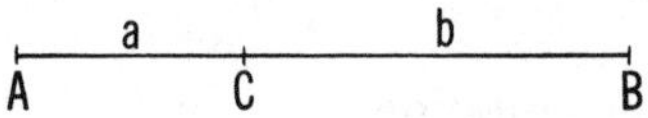

Figure 8

In 1509, Luca Pacioli published a book, *Divina Proportione*, devoted to a study of the number τ. The figures and drawings were made by Leonardo da Vinci. In this book Pacioli described thirteen interesting properties of τ.

The golden mean appears at many unexpected turns: in the pentagonal symmetry of certain flowers and marine animals, in the proportions of the human body, and so on. Man has employed the golden mean in the creative arts and in various aspects of contemporary design, especially in the printing and advertising crafts. For example, the majority of people considers that rectangle to be most pleasing, aesthetically, whose sides are in the approximate ratio 1 to τ. Witness the popularity of the 3×5 index card; the ratio 3 to 5 is approximately equal to the ratio 1 to τ.

In geometry, the golden mean is the key to the construction of the regular pentagon. The number τ occurs in connection with many mathematical games, and the convergents to τ also occur in connection with certain geometrical deceptions. The most familiar, perhaps, is the one involving a square 8 units by 8, which, as shown in Figure 9a, can seemingly be broken up and fitted together again to form a rectangle 5 by 13. The area of the square is $8 \cdot 8 = 64$, while that of the rectangle with what seem to be the same component parts is $5 \cdot 13 = 65$, so that somehow the area has been increased by 1 unit.

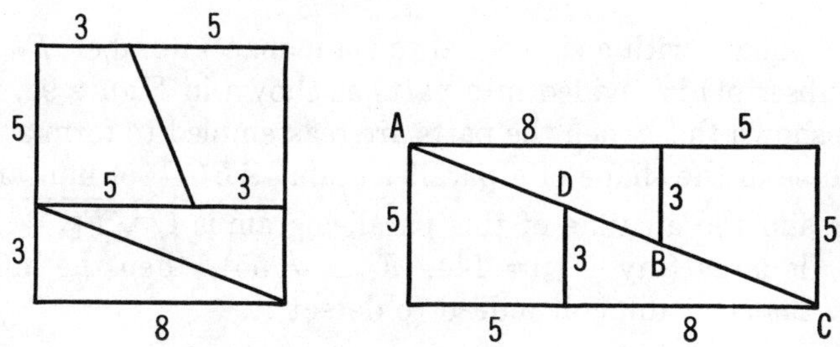

Figure 9a

This puzzle is based on the following facts: The convergents (3.15) have the property that the denominator of each is the numerator of the previous one. In particular,

$$\frac{p_5}{q_5} = \frac{8}{5}, \qquad \frac{p_6}{q_6} = \frac{13}{8}, \qquad q_6 = p_5 = 8.$$

Now consider the relation

$$p_n q_{n-1} - p_{n-1} q_n = (-1)^n$$

which in this case, for $n = 6$, becomes

$$13 \cdot 5 - 8 \cdot 8 = 1.$$

We have chosen p_6, q_5 as the dimensions of our rectangle, and p_5, q_6 ($p_5 = q_6$) as the dimensions of our square, and the above relation tells us that the areas of these figures differ by only one unit.

Actually, the points A, B, C, D do not lie on a straight line but are the vertices of a parallelogram $ABCD$ (for an exaggerated picture of the situation see Figure 9b) whose area is exactly equal to the "extra" unit of area. In case of the rectangle of Figure 9a, the obtuse angles ADC and ABC differ from straight angles by less than $1\frac{1}{4}°$.

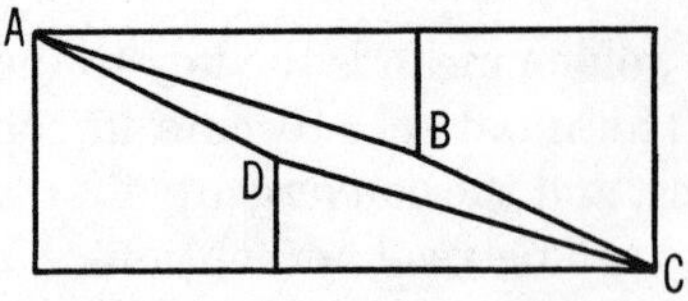

Figure 9b

More generally, if the Fibonacci numbers are defined by the relations

$$F_1 = 1, \quad F_2 = 1, \quad \text{and} \quad F_k = F_{k-2} + F_{k-1} \quad \text{for } k > 2,$$

and if a square with a side equal to a Fibonacci number F_{2n} (with even subscript) is divided into parts as shown in Figure 9c, then it can be shown that when the parts are reassembled to form a rectangle, a hole in the shape of a parallelogram $ABCD$ of unit area will appear and the altitude of this parallelogram is $1/\sqrt{F_{2n}^2 + F_{2n-2}^2}$. If F_{2n} is large (say $F_{2n} = 144$, $F_{2n-2} = 55$), then the hole is so narrow that it is difficult indeed to detect it.

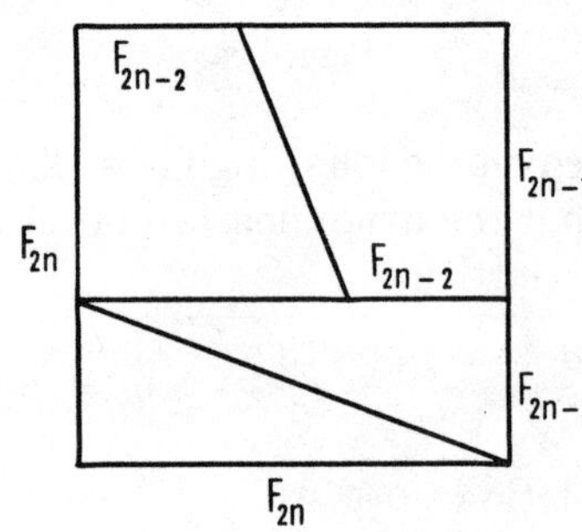

Figure 9c

3.11 A Method for Calculating Logarithms†

Daniel Shanks, in a journal devoted to numerical computations, [*Mathematical Tables and Other Aids to Computation*, Vol. 8, No. 45, April 1954, pp. 60–64], describes a method for calculating logarithms which is worth recording because of its adaptability to high-speed computing machines.

† This section is rather technical and may be omitted without loss of continuity.

To calculate the logarithm $\log_{b_0} b_1$ to the base b_0 of a number b_1 (where $1 < b_1 < b_0$) we compute two sequences:

$$b_2, b_3, b_4, \cdots$$

and the sequence of positive integers

$$n_1, n_2, n_3, \cdots,$$

where the numbers $n_1, b_2, n_2, b_3, \cdots$ are determined by means of the relations

$$\begin{array}{ll} b_1^{n_1} < b_0 < b_1^{n_1+1}, & b_2 = \dfrac{b_0}{b_1^{n_1}}, \\ b_2^{n_2} < b_1 < b_2^{n_2+1}, & b_3 = \dfrac{b_1}{b_2^{n_2}}, \\ \cdots\cdots\cdots\cdots, & \cdots\cdots\cdots, \\ b_k^{n_k} < b_{k-1} < b_k^{n_k+1}, & b_{k+1} = \dfrac{b_{k-1}}{b_k^{n_k}}, \\ \cdots\cdots\cdots\cdots, & \cdots\cdots\cdots. \end{array}$$

Thus, we first find an integer n_1 such that

$$b_1^{n_1} < b_0 < b_1^{n_1+1}.$$

This shows that

$$b_0 = b_1^{n_1+\frac{1}{x_1}}, \tag{3.17}$$

where $1/x_1 < 1$; we then calculate

$$b_2 = \frac{b_0}{b_1^{n_1}} \tag{3.18}$$

and determine an integer n_2 for which

$$b_2^{n_2} < b_1 < b_2^{n_2+1}.$$

If n_2 is such an integer, then

$$b_1 = b_2^{n_2+\frac{1}{x_2}}, \qquad x_2 > 1. \tag{3.19}$$

The procedure is now continued. Calculate

$$b_3 = \frac{b_1}{b_2^{n_2}}$$

and find an integer n_3 such that

$$b_3^{n_3} < b_2 < b_3^{n_3+1},$$

whence

$$b_2 = b_3^{n_3+\frac{1}{x_3}}, \qquad x_3 > 1,$$

and so on.

To see that we are actually calculating $\log_{b_0} b_1$ notice that from equations (3.17) and (3.18) we have

$$b_2 = b_0 b_1^{-n_1} = b_1^{n_1+\frac{1}{x_1}} b_1^{-n_1} = b_1^{\frac{1}{x_1}},$$

or

$$b_1 = b_2^{x_1}.$$

On the other hand, from (3.19),

$$b_1 = b_2^{n_2+\frac{1}{x_2}}$$

and hence we can write

$$x_1 = n_2 + \frac{1}{x_2}\cdot$$

Similarly we can show that

$$x_2 = n_3 + \frac{1}{x_3},$$

and so on. Solving equation (3.17) for b_1 and using these results we have

$$b_1 = b_0^{\frac{1}{n_1+\frac{1}{x_1}}} = b_0^{\frac{1}{n_1+\frac{1}{n_2+\frac{1}{x_2}}}}$$

$$= b_0^{\frac{1}{n_1+\frac{1}{n_2+\frac{1}{n_3+\ddots}}}},$$

and so, by the definition of a logarithm,

$$\log_{b_0} b_1 = \cfrac{1}{n_1 + \cfrac{1}{n_2 + \cfrac{1}{n_3 + \ddots}}}\cdot$$

EXAMPLE. Calculate $\log_{10} 2$.

SOLUTION. With $b_0 = 10$, $b_1 = 2$, we find that

$$2^3 < 10 < 2^4,$$

so that $n_1 = 3$ and $b_2 = 10/2^3 = 1.25$. Using a table of powers, we see that

$$(1.25)^3 < 2 < (1.25)^4.$$

Thus $n_2 = 3$ and $b_3 = 2/(1.25)^3 = 1.024$. Subsequent calculations become more difficult but can easily be done with the aid of a desk calculator. The paper by Shanks gives the following results:

$$\begin{array}{ll} b_1 = 2 & n_1 = 3 \\ b_2 = 1.25 & n_2 = 3 \\ b_3 = 1.024 & n_3 = 9 \\ b_4 = 1.009741958 & n_4 = 2 \\ b_5 = 1.004336279 & n_5 = 2 \\ \cdots\cdots\cdots\cdots\cdots & \cdots\cdots. \end{array}$$

This shows that

$$\log 2 = \frac{1}{3+}\,\frac{1}{3+}\,\frac{1}{9+}\,\frac{1}{2+}\,\frac{1}{2+}\cdots = [0, 3, 3, 9, 2, 2, \cdots].$$

Next we calculate the convergents:

i	-1	0	1	2	3	4	5	6
a_i			0	3	3	9	2	2
p_i	0	1	0	1	3	28	59	146
q_i	1	0	1	3	10	93	196	485
c_i			0	$\frac{1}{3}$	$\frac{3}{10}$	$\frac{28}{93}$	$\frac{59}{196}$	$\frac{146}{485}$

The convergent c_6 gives the approximation 0.30103093; the value of log 2 to 11 places is 0.30102999566. It can be shown that, in general, each convergent approximates log 2 to one more correct decimal place than does the previous convergent.

CHAPTER FOUR

Periodic Continued Fractions

4.1 Introduction

Our study so far has shown that rational numbers have finite continued fraction expansions, and that irrational numbers have non-terminating, or infinite, expansions.

In Chapter 3 we dealt mainly with the expansion of *quadratic irrationals*, or *quadratic surds*, i.e., with irrational numbers of the form

$$\frac{P \pm \sqrt{D}}{Q},$$

where P, Q, D are integers and where D is positive and not a perfect square. In all the examples considered, the expansions of such numbers were either purely periodic, like the expansion of $\frac{1}{3}(1 + \sqrt{10})$ below, or they were periodic from some point onward. For example,

$$\sqrt{2} = [1, 2, 2, 2, \cdots] = [1, \overline{2}],$$

$$\sqrt{19} = [4, \overline{2, 1, 3, 1, 2, 8}],$$

$$\frac{1 + \sqrt{10}}{3} = [1, 2, 1, 1, 2, 1, \cdots] = [\overline{1, 2, 1}],$$

where, as before, the bar over the partial quotients indicates those numbers which are repeated indefinitely. It is not hard to show that *any purely periodic continued fraction, or any fraction which is periodic from some point onward, represents a quadratic irrational.* The more difficult theorem, that *any quadratic irrational has a continued fraction expansion which is periodic after a certain stage,* was first proved by Lagrange in 1770. The aim of this chapter is the presentation of the proofs of these theorems. This will be accomplished in several stages.

First it will be shown that a purely periodic continued fraction represents a quadratic irrational of a special kind, called a *reduced* quadratic irrational; an example is presented at the beginning of Section 4.2 and is followed by the proof for the general case.

Section 4.3 furnishes a more detailed discussion of quadratic irrationals, and Section 4.4 supplies a deeper study of reduced quadratic irrationals. These sections contain the tools necessary for proving, in Section 4.5, that any reduced quadratic irrational has a purely periodic continued fraction expansion. This is followed by the proof of Lagrange's theorem which states that the continued fraction expansion of any quadratic irrational is periodic from some point on, and, conversely, every periodic continued fraction represents a quadratic irrational.

The chapter will end with a brief discussion of the indeterminate equation

$$x^2 - Ny^2 = 1, \tag{4.1}$$

where x and y are unknown integers, and where N is a given integer not a perfect square. In 1657 Fermat stated that equation (4.1) has infinitely many solutions, but he did not supply the proof.† Lord Brouncker in the same year gave a systematic method for solving the equation. The first complete discussion of (4.1) was given by Lagrange about 1766. Commonly, equation (4.1) is known as *Pell's equation;* but this is unjustified since Pell did not make any independent contribution to the subject.‡ Many authors refer to the equation as Fermat's equation.

References to indeterminate equations of the Pell type occur throughout the history of mathematics. The most interesting

† Actually proposed by Fermat as a challenge to English mathematicians of the time. For a complete history of the subject see Dickson [4, vol. 2, p. 341].

‡ John Pell (1611–1685) was a great teacher and scholar. Admitted to Trinity College, Cambridge, at the age of thirteen, Pell had mastered eight languages before he was twenty. He was professor of mathematics at Amsterdam (1643–1646), at Breda (1646–1652), and he was Cromwell's representative in Switzerland (1654–1658). He was elected a fellow of the Royal Society in 1663.

example arises in connection with the so-called "cattle problem" of Archimedes.† The solution of this problem contains eight unknowns (each representing the number of cattle of various kinds) which satisfy certain equations and conditions. The problem can be reduced to the equation

$$x^2 - 4729494y^2 = 1,$$

whose smallest solution involves numbers x and y with 45 and 41 digits respectively. The smallest solution of the cattle problem corresponding to these values of x and y consists of numbers with hundreds of thousands of digits. There is no evidence that the ancients came anywhere near to the solution of the problem. In fact some historians doubt that the problem had any connection with Archimedes, while others are convinced that it was propounded by Archimedes to Eratosthenes. See Heath [6, p. 121], Dickson [4, vol. 2, p. 342].

4.2 Purely Periodic Continued Fractions

Certain continued fractions, like

$$\sqrt{11} = [3, 3, 6, 3, 6, \cdots] = [3, \overline{3, 6}],$$

are periodic only after a certain stage. Others, like

$$\sqrt{11} + 3 = [6, 3, 6, 3, 6, \cdots] = [\overline{6, 3}],$$

are periodic from the beginning on and are called *purely periodic* continued fractions. Numbers represented by purely periodic continued fractions are quadratic irrationals of a particular kind, and we shall now investigate how these numbers can be distinguished from other quadratic irrationals.

(a) *A numerical example.* Consider some purely periodic continued fraction, such as

$$\alpha = [3, 1, 2, 3, 1, 2, \cdots] = [\overline{3, 1, 2}].$$

We can write

$$\alpha = 3 + \frac{1}{1+} \frac{1}{2+} \frac{1}{\alpha}. \tag{4.2}$$

† For a statement of the cattle problem see *The World of Mathematics* by James R. Newman, New York: Simon and Schuster, 1956, pp. 197–198.

It is now necessary to recall a result studied in Section 3.7. There we showed that if

$$\alpha = a_1 + \frac{1}{a_2 +} \cdots \frac{1}{+ a_n +} \frac{1}{\alpha_{n+1}}, \tag{4.3}$$

where

$$\alpha_{n+1} = a_{n+1} + \frac{1}{a_{n+2} + \cdots}, \tag{4.4}$$

then

$$\alpha = \frac{\alpha_{n+1}p_n + p_{n-1}}{\alpha_{n+1}q_n + q_{n-1}}, \tag{4.5}$$

where p_{n-1}/q_{n-1} and p_n/q_n are the convergents corresponding, respectively, to the partial quotients a_{n-1} and a_n. In effect, (4.5) shows that we can treat (4.3) as though it were a finite continued fraction, and that in calculating α we can regard α_{n+1} as though it were a legitimate partial quotient.

In the case of a purely periodic continued fraction

$$\alpha = [\overline{a_1, a_2, \cdots, a_n}] = a_1 + \frac{1}{a_2 +} \frac{1}{a_3 +} \cdots \frac{1}{+ a_n +} \frac{1}{\alpha_{n+1}},$$

we see that

$$\alpha_{n+1} = a_1 + \frac{1}{a_2 +} \frac{1}{a_3 + \cdots} = \alpha,$$

and hence equation (4.5) shows that α can be calculated from the equation

$$\alpha = \frac{\alpha p_n + p_{n-1}}{\alpha q_n + q_{n-1}}. \tag{4.6}$$

We now apply (4.6) to the special case (4.2), using $a_1 = 3$, $a_2 = 1$, $a_3 = 2$, $\alpha = [\overline{3, 1, 2}]$. We form the table

i	-1	0	1	2	3	4
a_i			3	1	2	α
p_i	0	1	3	4	11	$11\alpha + 4$
q_i	1	0	1	1	3	$3\alpha + 1$

Hence, we obtain

$$\alpha = \frac{\alpha p_3 + p_2}{\alpha q_3 + q_2} = \frac{11\alpha + 4}{3\alpha + 1}.$$

This leads to the quadratic equation

$$3\alpha^2 - 10\alpha - 4 = 0, \tag{4.7}$$

which is the same equation we would have obtained had we worked with equation (4.2).

We now consider the number β obtained from α by *reversing the period*, that is, the number

$$\beta = [\overline{2, 1, 3}] = 2 + \frac{1}{1+}\,\frac{1}{3+}\,\frac{1}{\beta}.$$

Applying (4.6) to β, we get

$$\beta = \frac{11\beta + 3}{4\beta + 1}; \tag{4.8}$$

this leads to the quadratic equation

$$4\beta^2 - 10\beta - 3 = 0. \tag{4.9}$$

Equation (4.9) can be written in the form

$$3\left(-\frac{1}{\beta}\right)^2 - 10\left(-\frac{1}{\beta}\right) - 4 = 0. \tag{4.10}$$

Comparing (4.7) and (4.10) we see that the quadratic equation

$$3x^2 - 10x - 4 = 0 \tag{4.11}$$

has solutions $x = \alpha$ and $x = -1/\beta$. These roots cannot be equal since both α and β are positive, and so α and $-1/\beta$ have opposite signs. Moreover, $\beta > 1$, and so $-1 < -1/\beta < 0$. This shows that the quadratic equation (4.7), or (4.11), has the positive root α and the negative root $\alpha' = -1/\beta$, where $-1 < \alpha' < 0$.

It is easy to check these results numerically. The quadratic formula shows that (4.7) has two roots,

$$\alpha = \frac{5 + \sqrt{37}}{3} \quad \text{and} \quad \alpha' = \frac{5 - \sqrt{37}}{3}.$$

The positive root β of (4.9) is

$$\beta = \frac{5 + \sqrt{37}}{4},$$

and hence

$$-\frac{1}{\beta} = \frac{-4}{5 + \sqrt{37}} = \frac{-4}{5 + \sqrt{37}} \cdot \frac{5 - \sqrt{37}}{5 - \sqrt{37}} = \frac{5 - \sqrt{37}}{3},$$

which shows that $-1/\beta$ is equal to α'. Moreover, to three decimal places, $\alpha = 3.694 > 1$, and $\alpha' = -0.361$, so that $-1 < \alpha' < 0$. The purely periodic continued fraction α is indeed a quadratic irrational.

(b) *The general case.* We shall now prove

THEOREM 4.1. *If* $a_1, a_2, \cdots, a_n$ *are positive integers, the purely periodic continued fraction*

$$\alpha = [\overline{a_1, a_2, \cdots, a_n}]$$

is greater than 1 *and is the positive root of a quadratic equation with integral coefficients. Moreover, if* $\beta = [\overline{a_n, a_{n-1}, \cdots, a_1}]$ *is the continued fraction for* α *with the period reversed, then* $-1/\beta = \alpha'$ *is the second, or conjugate root, of the quadratic equation satisfied by* α, *and, equally important,* α' *lies between* -1 *and* 0.

PROOF. We require two results stated in Problem 7 of Set 3, page 26, namely that if

$$\frac{p_n}{q_n} = [a_1, a_2, \cdots, a_{n-1}, a_n], \tag{4.12}$$

then

$$\frac{p_n}{p_{n-1}} = [a_n, a_{n-1}, \cdots, a_2, a_1] = \frac{p_n'}{q_n'}, \tag{4.13}$$

and

$$\frac{q_n}{q_{n-1}} = [a_n, a_{n-1}, \cdots, a_3, a_2] = \frac{p_{n-1}'}{q_{n-1}'}, \tag{4.14}$$

where p_n'/q_n' and p_{n-1}'/q_{n-1}' represent, respectively, the nth and $(n-1)$st convergents of the continued fraction $[a_n, a_{n-1}, \cdots, a_2, a_1]$.

Since convergents are in their lowest terms, it follows that

$$(4.15)\qquad \begin{aligned} p_n' &= p_n, & p_{n-1}' &= q_n,\\ q_n' &= p_{n-1}, & q_{n-1}' &= q_{n-1}. \end{aligned}$$

Since α is purely periodic we can write it in the form

$$\alpha = a_1 + \frac{1}{a_2 +} \cdots \frac{1}{+ a_n} \frac{1}{+ \alpha},$$

and, according to (4.6), in the form

$$(4.16)\qquad \alpha = \frac{\alpha p_n + p_{n-1}}{\alpha q_n + q_{n-1}},$$

where p_n/q_n and p_{n-1}/q_{n-1} are defined, respectively, as the nth and $(n-1)$st convergents of $[a_1, a_2, \cdots, a_n]$. Equation (4.16) is equivalent to the quadratic equation

$$(4.17)\qquad q_n\alpha^2 - (p_n - q_{n-1})\alpha - p_{n-1} = 0.$$

Reversing the period in α, we obtain

$$\beta = a_n + \frac{1}{a_{n-1} +} \cdots \frac{1}{+ a_1} \frac{1}{+ \beta},$$

and again, according to (4.6), we see that

$$(4.18)\qquad \beta = \frac{\beta p_n' + p_{n-1}'}{\beta q_n' + q_{n-1}'},$$

where p_n'/q_n' and p_{n-1}'/q_{n-1}' are, respectively, the nth and $(n-1)$st convergents to $[a_n, a_{n-1}, \ldots, a_1]$. Using the results stated in (4.15) we can replace (4.18) by

$$\beta = \frac{\beta p_n + q_n}{\beta p_{n-1} + q_{n-1}}$$

so that β satisfies the equation

$$p_{n-1}\beta^2 - (p_n - q_{n-1})\beta - q_n = 0,$$

which is equivalent to the equation

$$(4.19)\qquad q_n\left(-\frac{1}{\beta}\right)^2 - (p_n - q_{n-1})\left(-\frac{1}{\beta}\right) - p_{n-1} = 0.$$

Comparing equations (4.17) and (4.19), we conclude that the quadratic equation

$$q_n x^2 - (p_n - q_{n-1})x - p_{n-1} = 0$$

has two roots: The root $x_1 = \alpha$, and the root $x_2 = -1/\beta$. Now, β stands for the purely periodic continued fraction $[\overline{a_n, a_{n-1}, \cdots, a_1}]$, where $a_n, a_{n-1}, \cdots, a_1$ are all positive integers; thus we have $\beta > 1$, $0 < 1/\beta < 1$, and so $-1 < -1/\beta < 0$. In other words, the root $\alpha' = -1/\beta$ lies between -1 and 0. This completes the proof.

The converse of Theorem 4.1 is also true (and will be proved in Section 4.5). This means that if $\alpha > 1$ is a quadratic irrational number, and hence satisfies a quadratic equation with integral coefficients, and if the second root α' of this quadratic equation lies between -1 and 0, then the continued fraction expansion of α is *purely periodic.* This remarkable fact was first proved by Galois in 1828, though the result was implicit in the earlier work of Lagrange. What is to be emphasized is that *these few conditions on* α *and* α' *completely characterize the numbers which have purely periodic continued fraction expansions.*

Simple recurring continued fractions may be grouped as follows:

(i) Fractions which have no *acyclic* (or non-repeating) part, such as

$$\alpha = [\overline{a_1, a_2, \cdots, a_n}].$$

(ii) Those with an acyclic part consisting of a single quotient a_1, such as

$$\alpha = [a_1, \overline{b_1, b_2, b_3, \cdots, b_n}].$$

(iii) Those with an acyclic part containing *at least* two quotients, such as

$$\alpha = [a_1, a_2, a_3, \overline{b_1, b_2, \cdots, b_n}].$$

We proved, for fractions of type (i), that α is a quadratic irrational which satisfies a quadratic equation with integral coefficients, whose second root α' lies between -1 and 0. In cases (ii) and (iii) it can also be proved that α is a quadratic irrational satisfying a quadratic equation with integral coefficients, but in case (ii) the second root α' of this quadratic equation is either less than -1 or greater than 0, while in case (iii) the second root is necessarily greater than 0. We will not prove these last two results.

Problem Set 14

1. If $\alpha = [\overline{2, 6}]$ and $\beta = [\overline{6, 2}]$,
   (a) verify numerically that $\alpha > 1$ and $\beta > 1$,
   (b) find the equation of which α is a root,
   (c) show that the other root, α', of this equation satisfies the relation $\alpha' = -1/\beta$, and that α' therefore lies between -1 and 0.

2. Verify numerically
   (a) that $\alpha = [1, \overline{2, 3}]$ satisfies an equation whose other root, α', does *not* lie between -1 and 0,
   (b) that $\gamma = [1, 2, \overline{3}]$ satisfies an equation whose other root, γ', is positive.

4.3 Quadratic Irrationals

In this section we shall be concerned mainly with numbers of the form

$$A + B\sqrt{D},$$

where A and B are arbitrary rational numbers, and where D is a fixed positive integer not a perfect square, so that $\sqrt{D}$, and hence also $A + B\sqrt{D}$, are irrational.

First we observe that, for an arbitrary but fixed positive integer D, not a perfect square, there is only one way of writing the number $A + B\sqrt{D}$, aside from trivial variations such as

$$\tfrac{3}{2} + \tfrac{1}{3}\sqrt{5} = \tfrac{6}{4} + \tfrac{2}{6}\sqrt{5}.$$

In other words,

$$A_1 + B_1\sqrt{D} = A_2 + B_2\sqrt{D}$$

if and only if $A_1 = A_2$ and $B_1 = B_2$. To prove this, write the above equality in the form

$$A_1 - A_2 = (B_2 - B_1)\sqrt{D};$$

if $B_2 \neq B_1$, then

$$\sqrt{D} = \frac{A_1 - A_2}{B_2 - B_1}$$

would be rational, contrary to assumption. Hence the assumption that $B_1 \neq B_2$ leads to a contradiction and we must conclude that $B_1 = B_2$, and therefore, $A_1 - A_2 = 0$ or $A_1 = A_2$.

Next, we claim that when numbers of this form are combined by any of the elementary operations of arithmetic (addition, subtraction, multiplication, division), the result is again of this form. We leave the proofs of these properties to the reader (see Problem 1 of Set 15), but call attention to the fact that in this connection, "numbers of the form $A + B\sqrt{D}$" include those for which $B = 0$, i.e. ordinary rational numbers. When we speak of quadratic irrationals, however, we shall assume $B \neq 0$, since otherwise the number under consideration would be rational.

We prove next that *every number* $x = A + B\sqrt{D}$, *where* A *and* $B \neq 0$ *are rational and* D *is a positive integer, not a perfect square, is the root of a quadratic equation* $ax^2 + bx + c = 0$, *where the coefficients* $a > 0$, b, c *are integers and where* $b^2 - 4ac > 0$. Clearly if $a = 0$, $x = -c/b$ would be rational and hence could not represent the irrational number $A + B\sqrt{D}$.

In order to prove the statement in italics we recall that any quadratic equation

$$ax^2 + bx + c = 0, \qquad a > 0,$$

has roots

$$x = r_1 = -\frac{b}{2a} + \frac{\sqrt{b^2 - 4ac}}{2a} = A + B\sqrt{D},$$

$$x = r_2 = -\frac{b}{2a} - \frac{\sqrt{b^2 - 4ac}}{2a} = A - B\sqrt{D},$$

where $D = b^2 - 4ac$, and consequently

$$r_1 + r_2 = -\frac{b}{a} = 2A,$$

$$r_1 r_2 = \frac{c}{a} = A^2 - B^2 D.$$

Hence, if $a \neq 0$, we can replace $ax^2 + bx + c = 0$ by

$$x^2 - \left(-\frac{b}{a}\right)x + \frac{c}{a} = 0,$$

or by

$$x^2 - 2Ax + (A^2 - B^2 D) = 0.$$

Conversely, we can verify by direct substitution that

$$x = A + B\sqrt{D}$$

(and $x = A - B\sqrt{D}$) satisfies this last equation:

$$(A \pm B\sqrt{D})^2 - 2A(A \pm B\sqrt{D}) + (A^2 - B^2D)$$
$$= A^2 \pm 2AB\sqrt{D} + B^2D - 2A^2 \mp 2AB\sqrt{D} + A^2 - B^2D = 0.$$

The equation $x^2 - 2Ax + (A^2 - B^2D) = 0$ satisfied by $A + B\sqrt{D}$ and $A - B\sqrt{D}$ need not have integral coefficients, but if we multiply through by a, the common denominator of the rational numbers $2A$ and $A^2 - B^2D$, we obtain the quadratic equation

$$ax^2 + bx + c = 0$$

where the three coefficients $a > 0$, $b = -2aA$, and $c = a(A^2 - B^2D)$ are integers.

Finally, the discriminant $b^2 - 4ac$ of this last equation is positive; for,

$$b^2 - 4ac = (-2aA)^2 - 4a^2(A^2 - B^2D) = 4a^2B^2D > 0,$$

since D was assumed to be positive. Observe also that $b^2 - 4ac$ is not a perfect square.

The above discussion leads us to a precise definition of a *quadratic irrational*, or *quadratic surd;* it is a number which satisfies a quadratic equation whose coefficients are integers and whose discriminant is positive but not a perfect square. The numbers $A + B\sqrt{D}$ we have been dealing with are therefore all quadratic surds according to this definition, provided $B \neq 0$.

A quadratic surd $A + B\sqrt{D}$, $B \neq 0$, *satisfies one and only one quadratic equation* $ax^2 + bx + c = 0$ *where* a, b, c *have no factors in common.* For, if $x = A + B\sqrt{D}$ were a root of

$$g_1(x) = a_1x^2 + b_1x + c_1 = 0,$$

and also of

$$g_2(x) = a_2x^2 + b_2x + c_2 = 0,$$

then it would also be a root of the equation

$$a_2g_1(x) - a_1g_2(x) = (a_2b_1 - a_1b_2)x + (a_2c_1 - a_1c_2) = 0.$$

Now if $a_2b_1 - a_1b_2 \neq 0$, then this would imply that

$$x = -\frac{a_2c_1 - a_1c_2}{a_2b_1 - a_1b_2}$$

is rational, contrary to the assumption that x is irrational. Hence in this case $x = A + B\sqrt{D}$ could not satisfy both equations. On the other hand, if $a_2b_1 - a_1b_2 = 0$ then the equation

$$(a_2b_1 - a_1b_2)x + (a_2c_1 - a_1c_2) = 0$$

implies that $a_2c_1 - a_1c_2 = 0$, and hence that

$$\frac{a_2}{a_1} = \frac{b_2}{b_1} = \frac{c_2}{c_1} = k,$$

so that $a_2 = ka_1$, $b_2 = kb_1$, $c_2 = kc_1$ and the two quadratic equations $g_1(x) = 0$ and $g_2(x) = 0$ are actually equivalent, one being merely a constant multiple of the other.

Every quadratic irrational

$$\alpha = A + B\sqrt{D}$$

has a *conjugate*

$$\alpha' = A - B\sqrt{D}$$

formed by merely changing the sign of the coefficient B of $\sqrt{D}$. This definition has a number of useful consequences:

1. If α satisfies the quadratic equation $ax^2 + bx + c = 0$, then α' also satisfies this equation. (Why?)

2. The conjugate of the conjugate of a quadratic irrational number α is α. This follows directly from the definition of a conjugate, or from consequence 1., because a quadratic equation has only two roots.

3. The conjugate of the sum, difference, product, or quotient of two quadratic surds α_1 and α_2 is equal, respectively, to the sum, difference, product, or quotient of their conjugates. In symbols, this means that

$$(\alpha_1 + \alpha_2)' = \alpha_1' + \alpha_2',$$

$$(\alpha_1 - \alpha_2)' = \alpha_1' - \alpha_2',$$

$$(\alpha_1 \cdot \alpha_2)' = \alpha_1' \cdot \alpha_2',$$

$$\left(\frac{\alpha_1}{\alpha_2}\right)' = \frac{\alpha_1'}{\alpha_2'}.$$

We prove the first assertion, leaving the rest as problems. Thus if

$$\alpha_1 = A_1 + B_1\sqrt{D} \qquad \text{and} \qquad \alpha_2 = A_2 + B_2\sqrt{D},$$

then the conjugate of the sum is

$$\begin{aligned}(\alpha_1 + \alpha_2)' &= [(A_1 + A_2) + (B_1 + B_2)\sqrt{D}]' \\ &= (A_1 + A_2) - (B_1 + B_2)\sqrt{D}.\end{aligned}$$

On the other hand, the sum of the conjugates is

$$\begin{aligned}\alpha_1' + \alpha_2' &= (A_1 + B_1\sqrt{D})' + (A_2 + B_2\sqrt{D})' \\ &= (A_1 - B_1\sqrt{D}) + (A_2 - B_2\sqrt{D}) \\ &= (A_1 + A_2) - (B_1 + B_2)\sqrt{D},\end{aligned}$$

and comparing the two results we see that

$$(\alpha_1 + \alpha_2)' = \alpha_1' + \alpha_2'.$$

Problem Set 15

1. Show that, if $\alpha_1 = A_1 + B_1\sqrt{D}$, $\alpha_2 = A_2 + B_2\sqrt{D}$ (where A_1, A_2, B_1, B_2 are rational and D is a positive integer, not a perfect square), then $\alpha_1 + \alpha_2$, $\alpha_1 - \alpha_2$, $\alpha_1 \cdot \alpha_2$, α_1/α_2 $(\alpha_2 \neq 0)$, can each be expressed in the form $A + B\sqrt{D}$ with rational A, B.

2. Using the same representation of α_1, α_2 as in Problem 1, and denoting the conjugate of α by α', show that

$$(\alpha_1 - \alpha_2)' = \alpha_1' - \alpha_2', \qquad (\alpha_1 \cdot \alpha_2)' = \alpha_1' \cdot \alpha_2', \qquad \text{and} \qquad \left(\frac{\alpha_1}{\alpha_2}\right)' = \frac{\alpha_1'}{\alpha_2'}.$$

3. If $A + B\sqrt{M} + C\sqrt{N} = 0$ and if A, B, C are rational and M, N are positive integers, not perfect squares, such that $\sqrt{M}/\sqrt{N}$ is not rational, prove that $A = B = C = 0$.

4.4 Reduced Quadratic Irrationals

The quadratic equation

$$a\alpha^2 + b\alpha + c = 0, \qquad a > 0,$$

where a, b, c are integers, has roots

$$\alpha = \frac{-b + \sqrt{b^2 - 4ac}}{2a} = \frac{P + \sqrt{D}}{Q}, \tag{4.20}$$

and

$$\alpha' = \frac{-b - \sqrt{b^2 - 4ac}}{2a} = \frac{P - \sqrt{D}}{Q}, \tag{4.21}$$

where

$$P = -b, \qquad D = b^2 - 4ac, \qquad Q = 2a > 0 \tag{4.22}$$

are integers. If we assume that $D > 0$ is not a perfect square, then the roots α and α' are quadratic surds of the form $A \pm B\sqrt{D}$, where $A = P/Q$ and $B = 1/Q$ are rational.

Under these assumptions *the quadratic irrational* α *given by* (4.20) *is said to be reduced if* α *is greater than* 1 *and if its conjugate* α', *given by* (4.21), *lies between* -1 *and* 0. It is important in what follows to find out more about the form and properties of reduced quadratic irrationals. Throughout the rest of this chapter, P, Q, D will be as defined by (4.22).

Suppose, then, that the value of α given by (4.20) is a reduced quadratic irrational, i.e., that

$$\alpha = \frac{P + \sqrt{D}}{Q} > 1, \qquad \text{and} \qquad -1 < \alpha' = \frac{P - \sqrt{D}}{Q} < 0.$$

The conditions $\alpha > 1$ and $\alpha' > -1$ imply that $\alpha + \alpha' > 0$, or

$$\frac{P + \sqrt{D}}{Q} + \frac{P - \sqrt{D}}{Q} = \frac{2P}{Q} > 0,$$

and since $Q > 0$, we conclude that $P > 0$. Also, from

$$\alpha' = \frac{P - \sqrt{D}}{Q} < 0 \qquad \text{and} \qquad Q > 0,$$

it follows that $P - \sqrt{D} < 0$, or that $0 < P < \sqrt{D}$. The inequality $\alpha > 1$ implies that $P + \sqrt{D} > Q$; and the inequality $\alpha' > -1$ shows that $P - \sqrt{D} > -Q$, or $\sqrt{D} - P < Q$. Finally we observe that

$$P^2 - D = (-b)^2 - (b^2 - 4ac) = 4ac = 2c \cdot Q.$$

We have shown that if α is a reduced quadratic irrational of the form (4.20), then the integers P, Q, D satisfy the conditions

$$0 < P < \sqrt{D}, \text{ and } \sqrt{D} - P < Q < \sqrt{D} + P < 2\sqrt{D}. \tag{4.23}$$

The reason for introducing the notion of reduced quadratic surds has not been explained. This idea, however, is a well-established concept in the theory of numbers, and is intimately related to the theory of reduced quadratic forms. For our purpose, the importance of the idea depends upon the fact that *for any given* D *there is only a finite number of reduced quadratic surds of the form* (4.20). This follows directly from the inequalities (4.23); for, once D is fixed, there is only a finite number of positive integers P and Q such that $P < \sqrt{D}$ and $Q < 2\sqrt{D}$.

Could it happen that there are no reduced quadratic surds of the form $(P + \sqrt{D})/Q$ associated with a given D? If so we might be talking about an empty set of reduced surds. *However, for any given* $D > 1$, *not a perfect square, there exists always at least one reduced quadratic surd associated with it, namely*

$$\alpha = \lambda + \sqrt{D}$$

where λ *is the largest integer less than* $\sqrt{D}$. With this determination of λ, $\lambda + \sqrt{D} = \alpha$ is clearly greater than 1, and its conjugate $\alpha' = \lambda - \sqrt{D}$ obeys $-1 < \alpha' < 0$. The quadratic equation satisfied by α and α' is

$$x^2 - 2\lambda x + \lambda^2 - D = 0.$$

It is necessary to have the following result: *If* α *is a reduced quadratic surd, it may be expressed in the form*

$$\alpha = a_1 + \frac{1}{\alpha_1},$$

where a_1 *is the largest integer less than* α, *and where* α_1 *is again a reduced quadratic surd.*

To establish this result, let the reduced quadratic surd α be the root

$$\alpha = \frac{-b + \sqrt{b^2 - 4ac}}{2a} = \frac{P + \sqrt{D}}{Q}$$

of the equation $ax^2 + bx + c = 0$, where a, b, c are integers, $a > 0$, $P = -b$, $Q = 2a$, and $D = b^2 - 4ac > 0$ not a perfect square; see (4.22). Write α in the form $\alpha = a_1 + 1/\alpha_1$, where a_1 is the greatest integer less than α. Clearly $\alpha = a_1 + 1/\alpha_1$ satisfies the quadratic equation

$$a\left(a_1 + \frac{1}{\alpha_1}\right)^2 + b\left(a_1 + \frac{1}{\alpha_1}\right) + c = 0,$$

or

$$(aa_1^2 + ba_1 + c)\alpha_1^2 + (2aa_1 + b)\alpha_1 + a = 0.$$

Solving for the positive root α_1, we obtain

$$\alpha_1 = \frac{P_1 + \sqrt{D_1}}{Q_1},$$

where

$$P_1 = 2aa_1 + b, \qquad Q_1 = -2(aa_1^2 + ba_1 + c),$$

and

$$D_1 = (2aa_1 + b)^2 - 4a(aa_1^2 + ba_1 + c) = b^2 - 4ac = D.$$

These expressions give us the explicit form of α_1. It is also clear that P_1, Q_1, and $D_1 = D$ are integers, and

$$\alpha_1 = \frac{P_1 + \sqrt{D}}{Q_1}$$

has the same irrational part $\sqrt{D}$ as α has.

It will now be shown that α_1 is a reduced quadratic surd. To this end, we recall that a_1 is the greatest integer less than α; therefore $0 < 1/\alpha_1 < 1$, so $\alpha_1 > 1$, as required, and it only remains to prove that $-1 < \alpha_1' < 0$. Solving the equation $\alpha = a_1 + (1/\alpha_1)$ for α_1 and taking the conjugate of the result (see page 99), we obtain

$$\alpha_1' = \left(\frac{1}{\alpha - a_1}\right)' = \frac{1}{\alpha' - a_1}.$$

Therefore

$$-\frac{1}{\alpha_1'} = a_1 - \alpha' > 1,$$

since $a_1 \geq 1$ and, by hypotheses, $-1 < \alpha' < 0$. If follows that $0 < -\alpha_1' < 1$, or $-1 < \alpha_1' < 0$. Thus α_1 is a reduced quadratric surd, and hence the inequalities (4.23) are automatically inherited by P_1, Q_1, and $D_1 = D$.

Finally, we prove that if α is a reduced quadratic irrational, then its *associate* $\beta = -1/\alpha'$ is also a reduced quadratic irrational; for, the inequalities $\alpha > 1$, $-1 < \alpha' < 0$ imply that $\beta > 1$, and that $\beta' = -1/\alpha$ lies between -1 and 0.

Problem Set 16

1. Show that, if $\alpha = \frac{1}{3}(5 + \sqrt{37})$ is expressed in the form $\alpha = a_1 + (1/\alpha_1)$, where a_1 is the largest integer less than α, then α_1 is a reduced quadratic irrational.

2. Show that the conditions (4.23) are necessary and sufficient conditions for α [defined by equation (4.20)] to be a reduced quadratic irrational, In other words, prove that conditions (4.23) imply that $1 < \alpha$ and $-1 < \alpha' < 0$.

3. Determine all the reduced quadratic irrationals of the form $(P + \sqrt{43})/Q$.

4.5 Converse of Theorem 4.1

We are now ready to prove

THEOREM 4.2 (CONVERSE OF THEOREM 4.1). *If α is a reduced quadratic irrational, so that $\alpha > 1$ is the root of a quadratic equation with integral coefficients whose conjugate root α' lies between -1 and 0, then the continued fraction for α is purely periodic.*

PROOF. We first investigate the actual expansion of α into a continued fraction; then we show that this expansion is necessarily purely periodic.

The first step is to express the reduced quadratic irrational α in the form

$$\alpha = \frac{P + \sqrt{D}}{Q} = a_1 + \frac{1}{\alpha_1}, \tag{4.24}$$

where a_1 is the largest integer less than α, and where

$$\alpha_1 = \frac{P_1 + \sqrt{D}}{Q_1} > 1,$$

is again a reduced quadratic irrational associated with D. This we established in Section 4.4.

Step (4.24) is the first step in converting α into a continued fraction. Repeating the process on α_1, we obtain

$$\alpha_1 = \frac{P_1 + \sqrt{D}}{Q_1} = a_2 + \frac{1}{\alpha_2},$$

where a_2 is the largest integer less than α_1, and

$$\alpha_2 = \frac{P_2 + \sqrt{D}}{Q_2} > 1$$

is a reduced quadratic irrational. At this stage we have

$$\alpha = a_1 + \frac{1}{\alpha_1},$$

$$\alpha_1 = a_2 + \frac{1}{\alpha_2},$$

where α, α_1, α_2 are reduced, and

$$\alpha = a_1 + \frac{1}{a_2 +} \frac{1}{\alpha_2}.$$

Continuing the process, we generate step-by-step a string of equations

$$\alpha_0 = a_1 + \frac{1}{\alpha_1},$$

$$\alpha_1 = a_2 + \frac{1}{\alpha_2},$$

$$\cdots\cdots\cdots$$

$$\alpha_{n-1} = a_n + \frac{1}{\alpha_n},$$

$$\cdots\cdots\cdots,$$

where $\alpha_0 = \alpha$, α_1, α_2, $\cdots$ are all reduced quadratic irrationals associated with D, and where

$$\alpha = a_1 + \frac{1}{a_2 +} \cdots \frac{1}{+ a_n + \cdots}.$$

Since α is irrational this process never comes to an end, and hence we seemingly are generating an infinite number of reduced surds

$\alpha_0, \alpha_1, \cdots, \alpha_n, \cdots$, all associated with D. But we proved in Section 4.4 that there can only be a finite number of reduced α_i's associated with a given D; therefore, we must arrive eventually at a reduced surd which has occurred before. Suppose, then, that in the sequence

$$\alpha_0, \alpha_1, \cdots, \alpha_{k-1}, \alpha_k, \cdots, \alpha_{l-1}, \alpha_l, \cdots \tag{4.25}$$

all the complete quotients $\alpha_0, \alpha_1, \cdots, \alpha_{l-1}$ are different, and that α_l is the first one whose value has occurred before, so that $\alpha_l = \alpha_k$, $0 \leq k < l$.

It is then possible to prove that:

(i) Once a complete quotient is repeated, all subsequent complete quotients are repeated; in other words, $\alpha_k = \alpha_l$ implies $\alpha_{k+1} = \alpha_{l+1}$, $\alpha_{k+2} = \alpha_{l+2}, \cdots$.

(ii) The very first complete quotient, $\alpha = \alpha_0$, is repeated; in other words, the sequence $\alpha = \alpha_0, \alpha_1, \alpha_2, \cdots$ is purely periodic.

To prove (i), we merely recall that

$$\alpha_k = a_{k+1} + \frac{1}{\alpha_{k+1}} = \alpha_l = a_{l+1} + \frac{1}{\alpha_{l+1}},$$

and, since a_{k+1} and a_{l+1} are the greatest integers less than $\alpha_k = \alpha_l$, we may conclude that $a_{k+1} = a_{l+1}$. It then follows that the reciprocals of α_{k+1} and α_{l+1} are equal and hence that $\alpha_{k+1} = \alpha_{l+1}$. This argument, when repeated, also yields $\alpha_{k+2} = \alpha_{l+2}$, $\alpha_{k+3} = \alpha_{l+3}, \cdots$.

To prove (ii) we shall show that $\alpha_k = \alpha_l$ for $0 < k < l$ implies $\alpha_{k-1} = \alpha_{l-1}$, $\alpha_{k-2} = \alpha_{l-2}$, $\cdots$, $\alpha_0 = \alpha_{l-k}$. For this purpose, we use the conjugates of the equal complete quotients α_k and α_l, obtaining $\alpha_k' = \alpha_l'$, from which it follows that

$$\beta_k = -\frac{1}{\alpha_k'} = -\frac{1}{\alpha_l'} = \beta_l. \tag{4.26}$$

Now if $k \neq 0$, we have

$$\alpha_{k-1} = a_k + \frac{1}{\alpha_k} \quad \text{and} \quad \alpha_{l-1} = a_l + \frac{1}{\alpha_l};$$

taking conjugates, we obtain

$$\alpha_{k-1}' = a_k + \frac{1}{\alpha_k'} \quad \text{and} \quad \alpha_{l-1}' = a_l + \frac{1}{\alpha_l'},$$

and hence

$$-\frac{1}{\alpha_k'} = a_k - \alpha_{k-1}' \qquad \text{and} \qquad -\frac{1}{\alpha_l'} = a_l - \alpha_{l-1}',$$

which is the same as saying

$$\beta_k = a_k + \frac{1}{\beta_{k-1}} \qquad \text{and} \qquad \beta_l = a_l + \frac{1}{\beta_{l-1}} \cdot \tag{4.27}$$

Since α_{k-1}, α_{l-1} are reduced, we have

$$-1 < \alpha_{k-1}' < 0 \qquad \text{and} \qquad -1 < \alpha_{l-1}' < 0$$

so that

$$0 < -\alpha_{k-1}' = \frac{1}{\beta_{k-1}} < 1, \qquad \text{and} \qquad 0 < -\alpha_{l-1}' = \frac{1}{\beta_{l-1}} < 1.$$

This shows that the a_k, a_l in (4.27) are the largest integers less than β_k, β_l, respectively; and since $\beta_k = \beta_l$, it follows that $a_k = a_l$ and hence, also, that

$$a_k + \frac{1}{\alpha_k} = a_l + \frac{1}{\alpha_l} \cdot \tag{4.28}$$

Since the left side of (4.28) is α_{k-1} and the right side is α_{l-1}, we have shown that $\alpha_k = \alpha_l$ implies $\alpha_{k-1} = \alpha_{l-1}$. Now if $k - 1 \neq 0$, i.e., if α_k is not the very first complete quotient, we may repeat this argument k times to prove that

$$\alpha_{k-2} = \alpha_{l-2}, \qquad \alpha_{k-3} = \alpha_{l-3}, \qquad \text{etc.},$$

until we arrive at the first α, and obtain

$$\alpha_{k-k} = \alpha_0 = \alpha_{l-k} = \alpha_s.$$

Thus, in expanding the reduced quadratic irrational α into a continued fraction we generate the string of equations

$$\alpha = a_1 + \frac{1}{\alpha_1},$$

$$\alpha_1 = a_2 + \frac{1}{\alpha_2},$$

$$\cdots\cdots\cdots\cdots$$

$$\alpha_{s-2} = a_{s-1} + \frac{1}{\alpha_{s-1}},$$

$$\alpha_{s-1} = a_s + \frac{1}{\alpha_s} = a_s + \frac{1}{\alpha},$$

where $\alpha, \alpha_1, \alpha_2, \cdots, \alpha_{s-1}$ are all different, and where $\alpha_s = \alpha$, and from this point on the α's repeat.

Since for every $\alpha_k > 1$, there exists exactly one biggest integer a_k less than α_k, it is clear that the sequence $a_1, a_2, \cdots, a_s$ will also repeat:

$$\alpha_s = a_{s+1} + \frac{1}{\alpha_{s+1}} = \alpha_0 = a_1 + \frac{1}{\alpha_1} \cdot$$

Therefore, the continued fraction for α has the form

$$\alpha = \overline{[a_1, a_2, \cdots, a_s]}$$

of a purely periodic continued fraction. This completes the proof of Theorem 4.2.

Before extending the proof to all quadratic irrationals (reduced and not reduced), we present a graphical illustration of the periodic character of the complete quotients $\alpha_1, \alpha_2, \cdots$ in the expressions

$$\alpha = a_1 + \frac{1}{\alpha_1}, \qquad \alpha_1 = a_2 + \frac{1}{\alpha_2}, \qquad \cdots, \qquad \alpha_k = a_{k+1} + \frac{1}{\alpha_{k+1}}, \qquad \cdots.$$

We shall define two functions, $F(x)$ and $G(x)$, such that F maps α_n into $1/\alpha_{n+1}$ and G maps $1/\alpha_{n+1}$ into its reciprocal, α_{n+1}. By first applying F to some α_n, and then G to $F(\alpha_n)$, we shall obtain α_{n+1}.

To define the function F, observe that

$$\frac{1}{\alpha_{k+1}} = \alpha_k - a_{k+1}$$

where a_{k+1} is the largest integer less than α_k. Let the symbol $\{x\}$ denote the largest integer less than x;† then we may write

$$\frac{1}{\alpha_{k+1}} = \alpha_k - \{\alpha_k\},$$

and we define the function F accordingly:

$$F(x) = x - \{x\}.$$

We now have a function which assigns, to every α_k, the reciprocal of the next α; that is,

$$F(\alpha_k) = \alpha_k - \{\alpha_k\} = \frac{1}{\alpha_{k+1}} \cdot$$

† The traditional notation for "largest integer less than x" is $[x]$; but since this conflicts with our notation for continued fractions, we have adopted the braces here.

Now, since the reciprocal of the reciprocal of a number is the number itself, the appropriate definition of G is simply

$$G(x) = \frac{1}{x} \qquad \text{so that} \qquad G\left(\frac{1}{\alpha_{k+1}}\right) = \alpha_{k+1}.$$

In other words,

$$G[F(\alpha_k)] = \alpha_{k+1}.$$

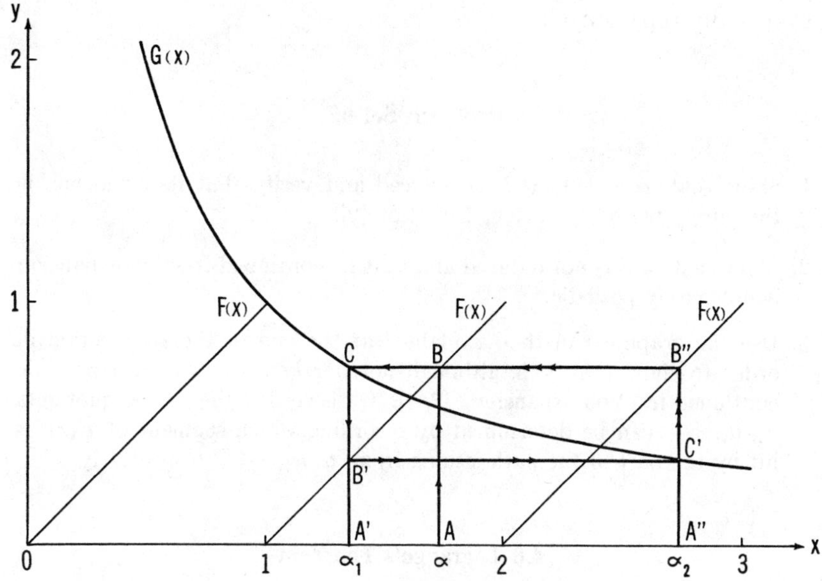

Figure 10

In order to apply this scheme graphically, plot the functions $F(x) = x - \{x\}$ and $G(x) = 1/x$ on the same graph paper; see Figure 10. The graph of $F(x)$ consists of the parallel line segments and the graph of $G(x)$, for positive x, consists of one branch of the equilateral hyperbola $y = 1/x$.

Let α be the given quadratic irrational. We locate it on the horizontal axis (point A) and find $F(\alpha) = 1/\alpha_1$ by measuring the vertical distance from A so the graph of $F(x)$ [i.e., to the point $F(\alpha) = B$]. We then find the point on the graph of $G(x)$ which has the same ordinate as the point B, namely $1/\alpha_1$; we call this point C. The projection of C onto the x-axis represents the value of α_1, because

$$G(\alpha_1) = \frac{1}{\alpha_1}\cdot$$

Starting with α_1, we now repeat this process, going from A' to B' to C'; the abscissa of C' represents the value of α_2.

The arrows in the figure indicate the paths that lead from each α to the next, a single arrow leading from α to α_1, a double arrow from α_1 to α_2, etc. If, in the course of our path, we are led to a point on the hyperbola which was already covered by the earlier part of the path, then there will be a repetition and the α_i are periodic. Conversely, if the α_i are periodic, then the path will eventually repeat itself.

Problem Set 17

1. Show that $\alpha = 1 + \sqrt{2}$ is reduced and verify that its expansion is the purely periodic continued fraction $[\overline{2}]$.

2. Show that $\sqrt{8}$ is not reduced and that its continued fraction expansion is not purely periodic.

3. Use the graphical method explained at the end of the last section in order to show that $\sqrt{5}$, although not purely periodic, has a periodic continued fraction expansion, $[2, \overline{4}]$. Observe that the partial quotients $a_1, a_2, \cdots$ can be determined by recording which segment of $F(x)$ is hit by the part of the path issuing from $\alpha, \alpha_1, \cdots$, respectively.

4.6 Lagrange's Theorem

THEOREM 4.3. *Any quadratic irrational number* α *has a continued fraction expansion which is periodic from some point onward.*

PROOF. The central idea of the proof is to show that when any quadratic irrational number α is developed into a continued fraction, eventually a *reduced* complete quotient α_{n+1} is reached, and from then on the fraction will be periodic by Theorem 4.2.

Let the expansion of α be

$$\alpha = a_1 + \frac{1}{a_2 +} \cdots \frac{1}{+ \alpha_{n+1}}.$$

Then, by equation (4.5) we know that

$$\alpha = \frac{\alpha_{n+1}p_n + p_{n-1}}{\alpha_{n+1}q_n + q_{n-1}},$$

where α and α_{n+1} are quadratic irrationals and $\alpha_{n+1} > 1$. Taking conjugates of both sides of this equation, we get

$$\alpha' = \frac{\alpha'_{n+1}p_n + p_{n-1}}{\alpha'_{n+1}q_n + q_{n-1}},$$

or, solving for α'_{n+1},

$$\alpha'_{n+1} = -\frac{\alpha' q_{n-1} - p_{n-1}}{\alpha' q_n - p_n}.$$

Factoring the numerator and the denominator, this gives

$$\alpha'_{n+1} = -\frac{q_{n-1}}{q_n}\left(\frac{\alpha' - \dfrac{p_{n-1}}{q_{n-1}}}{\alpha' - \dfrac{p_n}{q_n}}\right) \tag{4.29}$$

$$= -\frac{q_{n-1}}{q_n}\left(\frac{\alpha' - c_{n-1}}{\alpha' - c_n}\right),$$

where $c_{n-1} = p_{n-1}/q_{n-1}$ and $c_n = p_n/q_n$ are convergents to α. But from our study of convergents in Chapter 3 we know that as n increases indefinitely, both c_{n-1} and c_n tend to the limit α, and consequently

$$\frac{\alpha' - c_{n-1}}{\alpha' - c_n} \quad \text{tends to} \quad \frac{\alpha' - \alpha}{\alpha' - \alpha} = 1 \tag{4.30}$$

as n approaches infinity. We know also that the convergents c_n are alternately less than α and greater than α, and hence eventually, as n increases, the values of the fraction (4.30) will not only get closer and closer to 1, but they will be alternately slightly less than 1 and slightly greater than 1. We notice also that in (4.29) the numbers q_n and q_{n-1} are both positive integers and (see page 67) that $0 < q_{n-1} < q_n$, so that $q_{n-1}/q_n < 1$. Thus, once we have found a value of n which makes the fraction (4.30) slightly less than 1, the value of α'_{n+1} given by (4.29) will of necessity lie between -1 and 0. This proves that α_{n+1} is *reduced*; by Theorem 4.2 the continued fraction for α will be periodic from there on. Thus Lagrange's theorem has been proved.

Problem Set 18

1. Show that $\alpha = \frac{1}{9}(8 + \sqrt{37})$ is not reduced, but that, if

$$\alpha = a_1 + \frac{1}{a_2 +} \cdots \frac{1}{+ a_n +} \frac{1}{\alpha_{n+1}},$$

we eventually come to an α_{n+1} which is reduced, and verify that the expansion is periodic from then on.

4.7 The Continued Fraction for $\sqrt{N}$

If $N > 0$ is an integer which is not a perfect square, the continued fraction for $\sqrt{N}$ has an interesting form. First notice that $\sqrt{N}$ is greater than 1, and hence its conjugate $-\sqrt{N}$ cannot lie between -1 and 0, so $\sqrt{N}$ is *not reduced*, and its expansion

$$\sqrt{N} = a_1 + \frac{1}{a_2 +} \cdots \frac{1}{+ a_{n-1} +} \frac{1}{a_n +} \cdots \tag{4.31}$$

cannot be purely periodic. On the other hand, since a_1 is the largest integer less than $\sqrt{N}$, the number $\sqrt{N} + a_1$ is greater than 1, and its conjugate, $-\sqrt{N} + a_1$, does lie between -1 and 0, so $\sqrt{N} + a_1$ is reduced. Adding a_1 to both sides of (4.31) we get

$$\sqrt{N} + a_1 = 2a_1 + \frac{1}{a_2 +} \frac{1}{a_3 +} \cdots,$$

and since this expansion is purely periodic it must have the form

$$\begin{aligned} \alpha &= \sqrt{N} + a_1 \\ &= 2a_1 + \frac{1}{a_2 +} \frac{1}{a_3 +} \cdots \frac{1}{+ a_n +} \frac{1}{2a_1 +} \frac{1}{a_2 +} \cdots. \end{aligned} \tag{4.32}$$

Consequently, the expansion for $\sqrt{N}$ is

$$\begin{aligned} \sqrt{N} &= a_1 + \frac{1}{a_2 +} \frac{1}{a_3 +} \cdots \frac{1}{+ a_n +} \frac{1}{2a_1 +} \frac{1}{a_2 +} \cdots \\ &= [a_1, \overline{a_2, a_3, \cdots, a_n, 2a_1}], \end{aligned} \tag{4.33}$$

where the period starts after the first term and ends with the term $2a_1$. For example,

$$\sqrt{29} = [5, \overline{2, 1, 1, 2, 10}],$$

$$\sqrt{19} = [4, \overline{2, 1, 3, 1, 2, 8}].$$

Notice that, except for the term $2a_1$, the periodic part is *symmetrical.* The symmetrical part may or may not have a central term.

To investigate the symmetrical part, recall from Section 4.2 that if $\alpha' = -\sqrt{N} + a_1$ is the conjugate of $\alpha = \sqrt{N} + a_1$, then the expansion of $-1/\alpha'$ is the same as that of α, but with the period reversed. Hence, reversing the period in (4.32), we obtain

$$(4.34) \qquad -\frac{1}{\alpha'} = \frac{1}{\sqrt{N} - a_1} = a_n + \frac{1}{a_{n-1} +} \cdots \frac{1}{+ a_2 +} \frac{1}{2a_1 + \cdots}.$$

On the other hand, we can obtain the expansion for $(\sqrt{N} - a_1)^{-1}$ quite easily from (4.33); subtracting a_1 from both sides of this equation yields

$$\sqrt{N} - a_1 = 0 + \frac{1}{a_2 +} \frac{1}{a_3 +} \cdots \frac{1}{+ a_n +} \frac{1}{2a_1 +} \frac{1}{a_2 + \cdots},$$

and the reciprocal of this expression is

$$(4.35) \qquad \frac{1}{\sqrt{N} - a_1} = a_2 + \frac{1}{a_3 +} \cdots \frac{1}{+ a_n +} \frac{1}{2a_1 +} \frac{1}{a_2 + \cdots}.$$

We know, however, that continued fraction expansions are unique; hence, comparing (4.34) and (4.35), we conclude that

$$a_n = a_2, \qquad a_{n-1} = a_3, \qquad \cdots, \qquad a_3 = a_{n-1}, \qquad a_2 = a_n.$$

It follows that the continued fraction for $\sqrt{N}$ necessarily has the form

$$\sqrt{N} = [a_1, \overline{a_2, a_3, a_4, \cdots, a_4, a_3, a_2, 2a_1}].$$

See Table 2, page 116, for additional examples.

4.8 Pell's Equation, $x^2 - Ny^2 = \pm 1$

At the beginning of this chapter we mentioned that the cattle problem of Archimedes reduced to the solution of the equation

$$x^2 - 4729494y^2 = 1.$$

In this section we shall discuss the solutions in integers x and y of the equation

$$(4.36) \qquad x^2 - Ny^2 = 1,$$

where $N > 0$ is a given integer, and where x and y are unknown integers whose values we are seeking. We assume that N is *not a perfect square;* otherwise the equation is of little interest, since the difference of two perfect squares is never equal to 1 except in the special cases $(\pm 1)^2 - 0^2$. (Why?)

The continued fraction expansion for $\sqrt{N}$ supplies all the equipment we need to solve Pell's equation $x^2 - Ny^2 = 1$, or $x^2 - Ny^2 = -1$, provided solutions exist. We know that

$$\begin{aligned}\sqrt{N} &= a_1 + \frac{1}{a_2 +} \cdots \frac{1}{+ a_n +} \frac{1}{2a_1 +} \frac{1}{a_2 +} \cdots \\ &= a_1 + \frac{1}{a_2 +} \cdots \frac{1}{+ a_n +} \frac{1}{\alpha_{n+1}},\end{aligned} \tag{4.37}$$

where

$$\alpha_{n+1} = 2a_1 + \frac{1}{a_2 + \cdots} = \sqrt{N} + a_1. \tag{4.38}$$

We again use the fact that

$$\sqrt{N} = \frac{\alpha_{n+1} p_n + p_{n-1}}{\alpha_{n+1} q_n + q_{n-1}}, \tag{4.39}$$

where p_{n-1}, q_{n-1}, p_n, and q_n are calculated from the two convergents $c_{n-1} = p_{n-1}/q_{n-1}$, $c_n = p_n/q_n$ which come immediately before the term $2a_1$ in (4.37). Replacing α_{n+1} in (4.39) by the right side of (4.38) yields

$$\sqrt{N} = \frac{(\sqrt{N} + a_1)p_n + p_{n-1}}{(\sqrt{N} + a_1)q_n + q_{n-1}};$$

then, multiplying both sides by the denominator, we get

$$\sqrt{N}(\sqrt{N} + a_1)q_n + q_{n-1}\sqrt{N} = (\sqrt{N} + a_1)p_n + p_{n-1},$$

which is equivalent to

$$Nq_n + (a_1 q_n + q_{n-1})\sqrt{N} = (a_1 p_n + p_{n-1}) + p_n \sqrt{N}.$$

Now this is an equation of the form $a + b\sqrt{N} = c + d\sqrt{N}$, where a, b, c, d are integers and $\sqrt{N}$ is irrational, and this implies that $a = c$ and $b = d$ (see Section 4.3). Hence the last equation requires that

$$Nq_n = a_1 p_n + p_{n-1},$$

and

$$a_1 q_n + q_{n-1} = p_n.$$

Solving these equations for p_{n-1} and q_{n-1} in terms of p_n and q_n, we find that

$$p_{n-1} = Nq_n - a_1 p_n, \quad q_{n-1} = p_n - a_1 q_n. \tag{4.40}$$

But from Theorem 1.4 we know that

$$p_n q_{n-1} - q_n p_{n-1} = (-1)^n,$$

and, with the values of p_{n-1} and q_{n-1} from (4.40), this equation has the form

$$p_n(p_n - a_1 q_n) - q_n(Nq_n - a_1 p_n) = (-1)^n;$$

that is,

$$p_n^2 - Nq_n^2 = (-1)^n. \tag{4.41}$$

If n is *even*, equation (4.41) becomes

$$p_n^2 - Nq_n^2 = (-1)^n = 1,$$

and hence a particular solution of Pell's equation $x^2 - Ny^2 = 1$ is

$$x_1 = p_n, \qquad y_1 = q_n.$$

If n is *odd*, then

$$p_n^2 - Nq_n^2 = (-1)^n = -1,$$

and

$$x_1 = p_n, \qquad y_1 = q_n$$

gives a particular solution of the equation $x^2 - Ny^2 = -1$.

If n is *odd* and we still desire a solution the equation $x^2 - Ny^2 = 1$, we move ahead to the *second period* in the expansion of $\sqrt{N}$, that is, out to the term a_n where it occurs for the second time. Notice that

$$\sqrt{N} = a_1 + \frac{1}{a_2 +} \cdots \frac{1}{+ a_n +} \frac{1}{2a_1 +} \cdots \frac{1}{+ a_n +} \frac{1}{2a_1 +} \cdots$$

$$= a_1 + \frac{1}{a_2 +} \cdots \frac{1}{+ a_n +} \frac{1}{a_{n+1} +} \cdots \frac{1}{+ a_{2n} +} \frac{1}{a_{2n+1} +} \cdots,$$

so that the term a_n, when it occurs again, is actually the term a_{2n}; then

$$p_{2n}^2 - Nq_{2n}^2 = (-1)^{2n} = 1,$$

and so

$$x_1 = p_{2n}, \qquad y_1 = q_{2n},$$

gives us again a particular solution of the equation $x^2 - Ny^2 = 1$.

TABLE 2

N	Continued Fraction for $\sqrt{N}$	x_1	y_1	$x_1^2 - Ny_1^2$
2	$[1, \overline{2}]$	1	1	-1
3	$[1, \overline{1, 2}]$	2	1	$+1$
5	$[2, \overline{4}]$	2	1	-1
6	$[2, \overline{2, 4}]$	5	2	$+1$
7	$[2, \overline{1, 1, 1, 4}]$	8	3	$+1$
8	$[2, \overline{1, 4}]$	3	1	$+1$
10	$[3, \overline{6}]$	3	1	-1
11	$[3, \overline{3, 6}]$	10	3	$+1$
12	$[3, \overline{2, 6}]$	7	2	$+1$
13	$[3, \overline{1, 1, 1, 1, 6}]$	18	5	-1
14	$[3, \overline{1, 2, 1, 6}]$	15	4	$+1$
15	$[3, \overline{1, 6}]$	4	1	$+1$
17	$[4, \overline{8}]$	4	1	-1
18	$[4, \overline{4, 8}]$	17	4	$+1$
19	$[4, \overline{2, 1, 3, 1, 2, 8}]$	170	39	$+1$
20	$[4, \overline{2, 8}]$	9	2	$+1$
21	$[4, \overline{1, 1, 2, 1, 1, 8}]$	55	12	$+1$
22	$[4, \overline{1, 2, 4, 2, 1, 8}]$	197	42	$+1$
23	$[4, \overline{1, 3, 1, 8}]$	24	5	$+1$
24	$[4, \overline{1, 8}]$	5	1	$+1$
26	$[5, \overline{10}]$	5	1	-1
27	$[5, \overline{5, 10}]$	26	5	$+1$
28	$[5, \overline{3, 2, 3, 10}]$	127	24	$+1$
29	$[5, \overline{2, 1, 1, 2, 10}]$	70	13	-1
30	$[5, \overline{2, 10}]$	11	2	$+1$
31	$[5, \overline{1, 1, 3, 5, 3, 1, 1, 10}]$	1520	273	$+1$
32	$[5, \overline{1, 1, 1, 10}]$	17	3	$+1$
33	$[5, \overline{1, 2, 1, 10}]$	23	4	$+1$
34	$[5, \overline{1, 4, 1, 10}]$	35	6	$+1$
35	$[5, \overline{1, 10}]$	6	1	$+1$
37	$[6, \overline{12}]$	6	1	-1
38	$[6, \overline{6, 12}]$	37	6	$+1$
39	$[6, \overline{4, 12}]$	25	4	$+1$
40	$[6, \overline{3, 12}]$	19	3	$+1$

The above analysis shows that we can always find particular solutions of the equation

$$x^2 - Ny^2 = 1,$$

and sometimes particular solutions of the equation $x^2 - Ny^2 = -1$. Not all equations of the form $x^2 - Ny^2 = -1$ can be solved. For example, it can be proved (see Appendix I at the end of this book) that the equation $x^2 - 3y^2 = -1$ has no integral solutions. Here we shall confine our examples to equations that have solutions.

EXAMPLE 1. Find a particular solution of the equation $x^2 - 21y^2 = 1$.

SOLUTION. Here $N = 21$, and the continued fraction expansion given in Table 2 is

$$\sqrt{21} = [4, \overline{1, 1, 2, 1, 1, 8}] = [a_1, \overline{a_2, a_3, a_4, a_5, a_6, 2a_1}],$$

which shows that $a_n = a_6$, so that $n = 6$, an even number. A calculation shows that $c_6 = \frac{55}{12}$, so that

$$x_1 = p_6 = 55, \qquad y_1 = q_6 = 12,$$

and

$$x_1^2 - 21y_1^2 = 55^2 - 21 \cdot 12^2 = 3025 - 3024 = 1;$$

hence $x_1 = 55$, $y_1 = 12$ is a particular solution of the given equation.

EXAMPLE 2. Find a particular solution of the equation $x^2 - 29y^2 = 1$.

SOLUTION. The expansion of $\sqrt{29}$ is

$$\sqrt{29} = [5, \overline{2, 1, 1, 2, 10}] = [a_1, \overline{a_2, a_3, a_4, a_5, 2a_1}],$$

so that $n = 5$, an odd number. The first five convergents are

$$\frac{5}{1}, \quad \frac{11}{2}, \quad \frac{16}{3}, \quad \frac{27}{5}, \quad \frac{70}{13} = \frac{p_5}{q_5}.$$

But $x_1 = p_5 = 70$, $y_1 = q_5 = 13$, give $x^2 - 29y^2$ the value $70^2 - 29 \cdot 13^2 = -1$ and not $+1$. Hence, we must move on to the next period. The next period gives the convergents

$$\frac{727}{135}, \quad \frac{1524}{283}, \quad \frac{2251}{418}, \quad \frac{3775}{701}, \quad \frac{9801}{1820} = \frac{p_{10}}{q_{10}},$$

and so, if we take

$$x_1 = 9801, \qquad y_1 = 1820,$$

we get

$$x_1^2 - 29y_1^2 = 96059601 - 96059600 = 1.$$

The solutions arrived at in Example 1 can be checked against Table 2. In this Table, opposite $N = 21$ we find the expansion of

$$\sqrt{N} = \sqrt{21} = [4, \overline{1, 1, 2, 1, 1, 8}],$$

and further to the right we find listed a solution $x_1 = 55$, $y_1 = 12$ of the equation $x^2 - 21y^2 = 1$.

Likewise we can check Example 2, for the Table shows that

$$\sqrt{N} = \sqrt{29} = [5, \overline{2, 1, 1, 2, 10}]$$

and gives a solution $x_1 = 70$, $y_1 = 13$ of the equation $x^2 - 29y^2 = -1$, which indicates that we have to move to the next period to obtain a solution of the equation $x^2 - 29y^2 = +1$.

Problem Set 19

1. Show that $x_1 = 8$, $y_1 = 3$ is a solution of the equation $x^2 - 7y^2 = 1$, as indicated in Table 2.

2. Show that $x_1 = 18$, $y_1 = 5$ is a solution of the equation $x^2 - 13y^2 = -1$, and proceed to the next period to find a solution of the equation $x^2 - 13y^2 = 1$.

4.9 How to Obtain Other Solutions of Pell's Equation

We have seen that Pell's equation $x^2 - Ny^2 = 1$, N a positive integer not a perfect square, can always be solved, but that not all equations of the form $x^2 - Ny^2 = -1$ have solutions. However, *if* either of these equations has solutions, *then* the method outlined in Section 4.8 will always produce the *least positive* (*minimal*) *solution;* that is, it will always produce the two smallest integers $x_1 > 0$, $y_1 > 0$ such that $x_1^2 - Ny_1^2 = 1$ or $x_1^2 - Ny_1^2 = -1$. Once the least positive solution has been obtained, we can systematically generate all the other positive solutions. These statements will not be proved. We shall state the main theorems involved and illustrate them by examples.

THEOREM 4.4. *If* (x_1, y_1) *is the least positive solution of* $x^2 - Ny^2 = 1$, *then all the other positive solutions* (x_n, y_n) *can be obtained from the equation*

$$x_n + y_n\sqrt{N} = (x_1 + y_1\sqrt{N})^n \tag{4.42}$$

by setting, in turn, $n = 1, 2, 3, \cdots$.

The values of x_n and y_n are obtained from (4.42) by expanding the term $(x_1 + y_1\sqrt{N})^n$ by the binomial theorem and equating the rational parts and purely irrational parts of the resulting equation. For example, if (x_1, y_1) is the least positive solution of $x^2 - Ny^2 = 1$, then the solution (x_2, y_2) can be found by taking $n = 2$ in (4.42). This gives

$$x_2 + y_2\sqrt{N} = (x_1 + y_1\sqrt{N})^2 = (x_1^2 + Ny_1^2) + (2x_1y_1)\sqrt{N},$$

so that $x_2 = x_1^2 + Ny_1^2$ and $y_2 = 2x_1y_1$. Using these values, a direct calculation shows that

$$\begin{aligned} x_2^2 - Ny_2^2 &= (x_1^2 + Ny_1^2)^2 - N(2x_1y_1)^2 \\ &= x_1^4 + 2Nx_1^2y_1^2 + N^2y_1^4 - 4Nx_1^2y_1^2 \\ &= x_1^4 - 2Nx_1^2y_1^2 + N^2y_1^4 \\ &= (x_1^2 - Ny_1^2)^2 = 1, \end{aligned}$$

since by assumption (x_1, y_1) is a solution of $x^2 - Ny^2 = 1$.

It is easy to show that if x_n, y_n are calculated by equation (4.42), then $x_n^2 - Ny_n^2 = 1$. We have, from (4.42),

$$x_n + y_n\sqrt{N} = (x_1 + y_1\sqrt{N})(x_1 + y_1\sqrt{N}) \cdots (x_1 + y_1\sqrt{N}),$$

where there are n factors in the expression on the right-hand side. Since the conjugate of a product is the product of the conjugates, this gives

$$x_n - y_n\sqrt{N} = (x_1 - y_1\sqrt{N})(x_1 - y_1\sqrt{N}) \cdots (x_1 - y_1\sqrt{N}),$$

or

$$(4.43) \qquad x_n - y_n\sqrt{N} = (x_1 - y_1\sqrt{N})^n.$$

Now we factor $x_n^2 - Ny_n^2$ and use (4.42) and (4.43):

$$\begin{aligned} x_n^2 - Ny_n^2 &= (x_n + y_n\sqrt{N})(x_n - y_n\sqrt{N}) \\ &= (x_1 + y_1\sqrt{N})^n(x_1 - y_1\sqrt{N})^n \\ &= (x_1^2 - Ny_1^2)^n = 1. \end{aligned}$$

Thus x_n and y_n are solutions of the equation $x^2 - Ny^2 = 1$.

EXAMPLE 1. In Example 1 of Section 4.8 we found that $x_1 = 55$ and $y_1 = 12$ is a solution (minimal) of the equation $x^2 - 21y^2 = 1$. A second solution (x_2, y_2) can be obtained by setting $n = 2$ in (4.42); this gives

$$\begin{aligned} x_2 + y_2\sqrt{21} &= (55 + 12\sqrt{21})^2 \\ &= 3025 + 1320\sqrt{21} + 3024 \\ &= 6049 + 1320\sqrt{21}, \end{aligned}$$

which implies that $x_2 = 6049$, $y_2 = 1320$. These values satisfy the equation $x^2 - 21y^2 = 1$, since

$$(6049)^2 - 21(1320)^2 = 36590401 - 36590400 = 1.$$

In general, the solutions of Pell's equation become large very fast.

EXAMPLE 2. Table 2 shows that $x_1 = 2$, $y_1 = 1$ is a solution of the equation $x^2 - 3y^2 = 1$. A second solution (x_2, y_2) is given by the equation

$$x_2 + y_2\sqrt{3} = (2 + 1\sqrt{3})^2 = 7 + 4\sqrt{3},$$

so that $x_3 = 7$, $y_3 = 4$, and $7^2 - 3 \cdot 4^2 = 1$. A third solution (x_3, y_3) is given by the equation

$$x_3 + y_3\sqrt{3} = (2 + 1\sqrt{3})^3 = 26 + 15\sqrt{3},$$

so that $x_3 = 26$, $y_3 = 15$. This is true since

$$(26)^2 - 3(15)^2 = 676 - 675 = 1.$$

The procedure can be continued.

THEOREM 4.5. *Assuming that* $x^2 - Ny^2 = -1$ *is solvable, let* (x_1, y_1) *be the least positive solution. Then all positive solutions* (x_n, y_n) *of* $x^2 - Ny^2 = -1$ *can be calculated from the equation*

$$x_n + y_n\sqrt{N} = (x_1 + y_1\sqrt{N})^n, \tag{4.44}$$

by setting $n = 1, 3, 5, 7, \cdots$. *Moreover, using the same values* x_1, y_1, *all positive solutions of* $x^2 - Ny^2 = 1$ *are given by*

$$x_n + y_n\sqrt{N} = (x_1 + y_1\sqrt{N})^n, \tag{4.45}$$

with $n = 2, 4, 6, \cdots$.

EXAMPLE 3. Table 2 shows that $x_1 = 3$, $y_1 = 1$ is the minimal solution of $x^2 - 10y^2 = -1$. A second solution is obtained from (4.44) by setting $n = 3$. We have

$$x_3 + y_3\sqrt{10} = (3 + 1\sqrt{10})^3 = 117 + 37\sqrt{10},$$

so that $x_3 = 117$, $y_3 = 37$; this is a solution since

$$(117)^2 - 10(37)^2 = 13689 - 13690 = -1.$$

If we take $n = 2$ in (4.45), we get

$$x_2 + y_2\sqrt{10} = (3 + 1\sqrt{10})^2 = 19 + 6\sqrt{10}.$$

This gives $x_2 = 19$, $y_2 = 6$, and $19^2 - 10 \cdot 6^2 = 1$ so that these values are solutions of $x^2 - 10y^2 = 1$.

In concluding this section, we remark that the study of the equation $x^2 - Ny^2 = 1$ is preliminary to the study of the most general equation of second degree in two unknowns, equations of the form

$$Ax^2 + Bxy + Cy^2 + Dx + Ey + F = 0,$$

where A, B, C, D, E and F are integers, and where x and y are the unknown integers. By means of certain substitutions for the variables x and y, the solutions of this equation (if they exist) can be made to depend upon the corresponding solutions of an equation of the form $x^2 - Ny^2 = M$. This involves an extensive study, and so we must be content with this introduction.

Problem Set 20

1. Table 2 indicates that $x_1 = 17$, $y_1 = 4$ is the minimal solution of the equation $x^2 - 18y^2 = 1$. Use Theorem 4.4 to find the next two solutions.

2. Table 2 shows that $x_1 = 18$, $y_1 = 5$ is the minimal solution of the equation $x^2 - 13y^2 = -1$. Use Theorem 4.5 to find the next solution. Also, find two solutions of the equation $x^2 - 13y^2 = 1$.

3. Consider the Pythagorean equation $x^2 + y^2 = z^2$; if m and n are integers, then the values

$$x = 2mn, \qquad y = m^2 - n^2, \qquad z = m^2 + n^2$$

will always give integral solutions of $x^2 + y^2 = z^2$ because of the identity

$$(2mn)^2 + (m^2 - n^2)^2 = (m^2 + n^2)^2.$$

We now propose the problem of finding right triangles with legs of lengths x and y, see Figure 11, so that x and y are consecutive integers. Then,

$$y - x = m^2 - n^2 - 2mn = (m - n)^2 - 2n^2 = \pm 1.$$

Let $m - n = u$, $n = v$, so that $m = u + n = u + v$. Now the problem is reduced to finding integral solutions of the equation

$$u^2 - 2v^2 = \pm 1.$$

Solve this equation and list the first four solutions of $x^2 + y^2 = z^2$ such that $y - x = \pm 1$.

4. Find sets of integers (x, y, z) for the sides of the right triangle of Figure 11 such that, as these integers increase, the angle θ between x and z approaches $60°$.

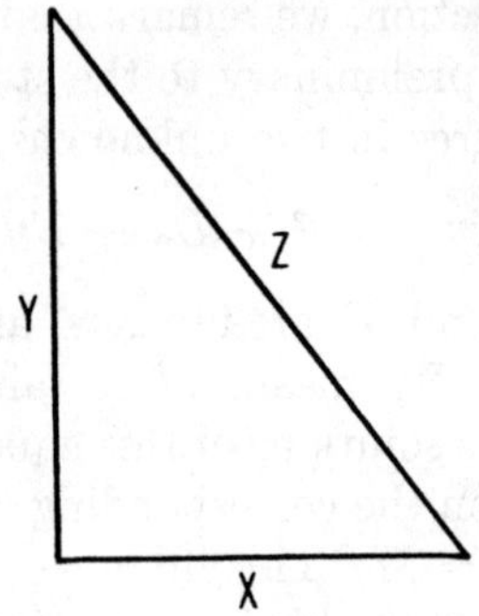

Figure 11

CHAPTER FIVE

Epilogue

5.1 Introduction

In this chapter we shall preview some results that can be studied once the first four chapters of this book have been mastered. We have already indicated that a complete study of Pell's equation $x^2 - Ny^2 = M$ could be undertaken, and would lead to the general solution, in integers, of the equation

$$Ax^2 + Bxy + Cy^2 + Dx + Ey + F = 0.$$

We shall concentrate now, however, on theorems related to the approximation of an irrational number by a rational fraction. Proofs of the theorems stated here and of many related theorems can be found in the books by Niven [8], and Hardy and Wright [5].

5.2 Statement of the Problem

Throughout this chapter let α be a given irrational number, and let p/q be a rational fraction, where p and q have no factors in common. It is clear that we can always find a rational fraction p/q, with positive q, as close as we please to α; in other words, if ϵ is any given number, however small, we can always find relatively prime integers p, q such that

$$\left|\alpha - \frac{p}{q}\right| < \epsilon. \tag{5.1}$$

But this is not the interesting point. What we should like to know is: Given α and ϵ in (5.1), how large must q be? Or, given α and q, how small can we make ϵ ?

We have already accomplished something along these lines. We proved in Chapter 3, Theorem 3.9, that if α is irrational, there exists an infinite number of rational fractions in lowest terms, of the form p/q, $q > 0$, such that

$$\left|\alpha - \frac{p}{q}\right| < \frac{1}{q^2}. \tag{5.2}$$

Any of the convergents p_1/q_1, p_2/q_2, $\cdots$, p_n/q_n, $\cdots$ to the continued fraction expansion of α can serve as the fraction p/q in (5.2).

It is possible to sharpen the inequality (5.2) as shown by the following theorem, stated here without proof.

THEOREM 5.1. *Of any two consecutive convergents* p_n/q_n *and* p_{n+1}/q_{n+1} *to the continued fraction expansion of* α, *at least one (call it* p/q*) satisfies the inequality*

$$\left|\alpha - \frac{p}{q}\right| < \frac{1}{2q^2}. \tag{5.3}$$

Moreover, the inequality (5.3) has this interesting feature: *If* α *is any irrational number, and if* p/q *is a rational fraction in lowest terms, with* $q \geq 1$, *such that*

$$\left|\alpha - \frac{p}{q}\right| < \frac{1}{2q^2},$$

then it can be proved that p/q *is necessarily one of the convergents of the simple continued fraction expansion of* α.

5.3 Hurwitz's Theorem

Inequality (5.3) immediately suggests the following question concerning still better approximations. Given an irrational number α, is there a number $k > 2$ such that the inequality

$$\left|\alpha - \frac{p}{q}\right| < \frac{1}{kq^2}, \qquad q \geq 1, \tag{5.4}$$

has infinitely many solutions p/q ? If so, then how large can k be?

It can be shown that, if the continued fraction expansion of α is $[a_1, a_2, \cdots, a_n, \cdots]$, and if p_n/q_n is the nth convergent, then

$$\left|\alpha - \frac{p_n}{q_n}\right| < \frac{1}{a_n q_n^2}\,; \tag{5.5}$$

hence we can get very good approximations to α if the numbers $a_1, a_2, \cdots, a_n, \cdots$ get large very fast. On the other hand if there are small numbers in the sequence $a_1, a_2, \cdots, a_n, \cdots$, no matter how far out we go, then the rational approximations p_n/q_n cannot be too good for small a_n.

From the point of view of approximation, the "simplest" numbers are the worst in the following sense: The "simplest" irrational number is

$$\xi = \frac{\sqrt{5} - 1}{2} = [0, 1, 1, \cdots] = [0, \bar{1}],$$

where each a_i has the smallest possible value. The convergents to ξ are the fractions

$$\frac{0}{1}, \quad \frac{1}{1}, \quad \frac{1}{2}, \quad \frac{2}{3}, \quad \frac{3}{5}, \quad \frac{5}{8}, \quad \cdots,$$

so that $q_{n-1} = p_n$ and

$$\frac{q_{n-1}}{q_n} = \frac{p_n}{q_n} \to \xi.$$

It can be shown that, for n very large, the expression

$$\left|\xi - \frac{p_n}{q_n}\right| = \left|\frac{\sqrt{5} - 1}{2} - \frac{p_n}{q_n}\right|$$

gets closer and closer to $1/\sqrt{5}\, q_n^2$.

These remarks suggest the truth of the following theorem, first proved by Hurwitz in 1891.

THEOREM 5.2. *Any irrational number α has an infinity of rational approximations p/q which satisfy the inequality*

$$\left|\alpha - \frac{p}{q}\right| < \frac{1}{\sqrt{5}\, q^2}\,, \qquad q \geq 1. \tag{5.6}$$

The number $\sqrt{5}$ is the best possible number; the theorem would become false if any larger number were substituted for $\sqrt{5}$.

By "false" we mean here that if $\sqrt{5}$ were replaced by any number $k > \sqrt{5}$, then there exists only a *finite* number of such rational approximations p/q to α, not an infinite number. Niven [8] gives an elementary proof that $\sqrt{5}$ is the best possible number in this sense.

One proof (by means of continued fractions) of Theorem 5.2 depends upon the fact that, in the continued fraction expansion of α, at least one of every three consecutive convergents beyond the first satisfies the inequality (5.6).

In his original proof of Theorem 5.2, Hurwitz did not use continued fractions; instead he based his proof on properties of certain fractions known as the *Farey sequences.* For any positive integer n, the sequence F_n is the set of rational numbers a/b with $0 \leq a \leq b \leq n$, $(a, b) = 1$, arranged in increasing order of magnitude. The first four sequences are:

$$F_1: \quad \frac{0}{1}, \quad \frac{1}{1},$$

$$F_2: \quad \frac{0}{1}, \quad \frac{1}{2}, \quad \frac{1}{1},$$

$$F_3: \quad \frac{0}{1}, \quad \frac{1}{3}, \quad \frac{1}{2}, \quad \frac{2}{3}, \quad \frac{1}{1},$$

$$F_4: \quad \frac{0}{1}, \quad \frac{1}{4}, \quad \frac{1}{3}, \quad \frac{1}{2}, \quad \frac{2}{3}, \quad \frac{3}{4}, \quad \frac{1}{1}.$$

These sequences have many useful properties; the one important for this study is: *If for any n, the irrational number $0 < \beta < 1$ lies between two consecutive fractions p/q, r/s of the sequence F_n, then at least one of the three ratios p/q, $(p + r)/(q + s)$, r/s can be used for x/y in the inequality*

$$\left|\beta - \frac{x}{y}\right| < \frac{1}{\sqrt{5}\, y^2}.$$

In order to make such an inequality valid also for an irrational number $\alpha > 1$, we let $\beta = \alpha - n$ where n is the greatest integer less than α. Substituting for β in the above inequality, we obtain

$$\left|\alpha - \left(n + \frac{x}{y}\right)\right| < \frac{1}{\sqrt{5}\, y^2} \quad \text{or} \quad \left|\alpha - \frac{x'}{y}\right| < \frac{1}{\sqrt{5}\, y^2},$$

where $x' = ny + x$. This is the central idea in Hurwitz's proof. For complete details see LeVeque [7].

Mathematicians are never content with a "best possible result", such as the constant $\sqrt{5}$ in Theorem 5.2. Such a statement always seems to stimulate further research. If a certain class of irrationals were ruled out, could this constant perhaps be replaced by a larger number? Indeed this can be done. The class of irrationals to be excluded consists of all numbers *equivalent* to the critical number $\xi = \frac{1}{2}(\sqrt{5} - 1)$ which forced us to accept $\sqrt{5}$ as the "best possible" constant in the inequality (5.6). We shall show that all numbers equivalent to ξ have the same periodic part at the end of their continued fraction expansions as ξ has, and are therefore just as hard to approximate.

DEFINITION: Here a number x is said to be *equivalent* to a number y (in symbols, $x \sim y$) if there are integers a, b, c, d satisfying the condition

$$ad - bc = \pm 1 \tag{5.7}$$

and such that x can be expressed in terms of y by the fraction

$$x = \frac{ay + b}{cy + d}. \tag{5.8}$$

For example, if $y = \sqrt{2}$ and $x = (2\sqrt{2} + 3)/(\sqrt{2} + 1)$, $x \sim y$ because $x = (a\sqrt{2} + b)/(c\sqrt{2} + d)$ with $a = 2$, $b = 3$, $c = 1$, $d = 1$, and $ad - bc = 2 - 3 = -1$. It is easy to see that the equivalence just defined has all the properties required of an equivalence relation, namely that it be

(i) *reflexive*, i.e., every x is equivalent to itself ($x \sim x$),
(ii) *symmetric*, i.e., if $x \sim y$, then $y \sim x$,
(iii) *transitive*, i.e., if $x \sim y$ and $y \sim z$, then $x \sim z$.

An equivalence relation divides the set of all numbers into *equivalence classes* in such a way that each number belongs to one and only one equivalence class.

Now, if a real number α has the continued fraction expansion

$$\alpha = [a_1, a_2, \cdots, a_n, \alpha_{n+1}],$$

it follows from

$$\alpha = \frac{\alpha_{n+1}p_n + p_{n-1}}{\alpha_{n+1}q_n + q_{n-1}}$$

and from $p_nq_{n-1} - p_{n-1}q_n = (-1)^n$ (see Theorem 1.4), that $\alpha \sim \alpha_{n+1}$ [cf. (5.7) and (5.8)]. Hence if α and β are any two real numbers with continued fraction expansions

$$\alpha = [a_1, a_2, \cdots, a_n, \alpha_{n+1}], \qquad \beta = [b_1, b_2, \cdots, b_m, \beta_{m+1}],$$

and if $\alpha_{n+1} = \beta_{m+1}$, then $\alpha \sim \alpha_{n+1} \sim \beta_{m+1} \sim \beta$, so $\alpha \sim \beta$. In particular any two rational numbers x and y are equivalent, for their expansions can always be written in the form

$$x = [a_1, a_2, \cdots, a_n, 1],$$
$$y = [b_1, b_2, \cdots, b_m, 1];$$

and since $1 \sim 1$, $x \sim y$.

The question as to when one irrational number is equivalent to another is answered by the following theorem, stated here without proof.

THEOREM 5.3. *Two irrational numbers* α *and* β *are equivalent if and only if*

$$\alpha = [a_1, a_2, \cdots, a_m, c_0, c_1, c_2, \ldots],$$
$$\beta = [b_1, b_2, \cdots, b_n, c_0, c_1, c_2, \ldots];$$

that is, if and only if the sequence of quotients in α *after the* mth *is the same as the sequence in* β *after the* nth.

Now let us return to Hurwitz's theorem. There are infinitely many irrational numbers equivalent to $\xi = \frac{1}{2}(\sqrt{5} - 1)$; let us suppose that each of these is expanded into a simple continued fraction. Then, by Theorem 5.3, from a certain place on, each of these expansions will contain the same sequence of quotients, c_0, c_1, $c_2, \cdots$, and hence all these equivalent irrationals play essentially the same role in Hurwitz's theorem as the number $\xi = \frac{1}{2}(\sqrt{5} - 1)$ does. It seems reasonable to guess that if we rule out the number ξ and all irrationals equivalent to it, then the constant $\sqrt{5}$ in Hurwitz's theorem could be replaced by a larger number. In fact the following theorem can be proved.

THEOREM 5.4. *Any irrational number* β *not equivalent to* $\xi = \frac{1}{2}(1 - \sqrt{5})$ *has an infinity of rational approximations* p/q *which satisfy the inequality*

$$\left|\beta - \frac{p}{q}\right| < \frac{1}{\sqrt{8}\, q^2}. \tag{5.9}$$

There is a chain of theorems similar to this one. For example, *if* β *is not equivalent to either* $\frac{1}{2}(\sqrt{5} - 1)$ *or* $\sqrt{2}$, *then the number* $\sqrt{8}$ *in* (5.9) *can be replaced by any number less than or equal to* $\sqrt{221}/5$.

Recently interest has been shown in "lop-sided" or unsymmetrical approximations to irrational numbers. For example, the following theorem was proved by B. Segre in 1946, and a very simple proof using Farey sequences was recently given by Ivan Niven.†

THEOREM 5.5. *For any real number* $r \geq 0$, *an irrational number* α *can be approximated by infinitely many rational fractions* p/q *in such a way that*

$$-\frac{1}{\sqrt{1+4r}\,q^2} < \frac{p}{q} - \alpha < \frac{r}{\sqrt{1+4r}\,q^2}.$$

When $r = 1$, this is Hurwitz's Theorem. For $r \neq 1$, notice that the lower bound is not just the negative of the upper bound, and the expression is unsymmetrical.

Using continued fractions, R. M. Robinson (1947) gave a proof of Segre's theorem, and also proved that given $\epsilon > 0$, the inequality

$$-\frac{1}{(\sqrt{5}-\epsilon)q^2} < \frac{p}{q} - \alpha < \frac{1}{(\sqrt{5}+1)q^2}$$

has infinitely many solutions. This result is interesting since it shows that one side of Hurwitz's inequality can be strengthened without essentially weakening the other.

5.4 Conclusion

Hurwitz's theorem is an example of a whole class of related theorems and problems studied under the general title of Diophantine approximations. The subject has a long history; yet there are still many challenging problems left to be solved. In recent years several new methods for solving problems in this field have been invented, but the study of continued fractions is, and probably will remain, the basic stepping stone for those wishing to explore this subject.

The field of Diophantine approximations by no means exhausts the avenues of exploration open to the interested student; this monograph can serve as the point of departure for further study of a variety of topics. One could, of course, go into the subject of continued fractions more deeply by reading such books as Perron [11].

† *On Asymmetric Diophantine Approximations*, The Michigan Math. Journal, vol. 9, No. 2, 1962, pp. 121–123.

Alternatively, there is the extension to analytic continued fractions (see Wall [14]), a subject initiated by Stieltjes and others; and there is the beautiful and closely related subject of the geometry of numbers, founded by Minkowski. For an introduction to the geometry of numbers, see Hardy and Wright [5], Chapters 3, 24.

Problem Set 21

1. Calculate the first six convergents to $\alpha = \frac{1}{3}(1 + \sqrt{10})$ and show that of every three of these consecutive convergents beyond the first, at least one satisfies Hurwitz's inequality (5.6).

2. Calculate the next row, F_5, of the Farey sequences given on page 126.

3. Locate $\alpha = \frac{1}{3}(\sqrt{10} - 2)$ between two successive elements p/q, r/s of the Farey sequence F_2 on page 126 and verify that at least one of the numbers p/q, $(p + r)/(q + s)$, r/s satisfies the inequality (5.6).

4. If $x = \frac{1}{2}(1 + \sqrt{5})$, show that $y = (-10x + 7)/(7x - 5)$ is equivalent to x. Expand both x and y into simple continued fractions and use these to give a numerical check on Theorem 5.3.

5. Prove that the equivalence relation defined on page 127 is (i) reflexive, (ii) symmetric, and (iii) transitive.

APPENDIX I

Proof That $x^2 - 3y^2 = -1$ Has No Integral Solutions

To show that the equation $x^2 - 3y^2 = -1$ is not solvable in integers x, y, we notice first that x and y cannot be both even or both odd. For, in the first case, if $x = 2x_1$, $y = 2y_1$ are both even integers, then

$$x^2 - 3y^2 = 4(x_1^2 - 3y_1^2)$$

is even and so could not be equal to -1. Similarly, in the second case, if $x = 2x_1 + 1$, $y = 2y_1 + 1$ are both odd integers, then

$$\begin{aligned} x^2 - 3y^2 &= (2x_1 + 1)^2 - 3(2y_1 + 1)^2 \\ &= 2(2x_1^2 - 6y_1^2 + 2x_1 - 6y_1 - 1) \end{aligned}$$

is also even (twice an integer) and again could not equal -1. Hence, if $x^2 - 3y^2 = -1$ is to have integral solutions, then we must have x even, y odd; or x odd, y even.

Suppose that x is even and y is odd, so that $x = 2x_1$, $y = 2y_1 + 1$. Then

$$(1) \qquad y^2 = 4y_1^2 + 4y_1 + 1 = 4y_1(y_1 + 1) + 1,$$

and since y_1 and $y_1 + 1$ are consecutive integers one of them must be even. So $y_1(y_1 + 1)$ is divisible by 2; hence $4y_1(y_1 + 1)$ is divisible by 8, and, from (1), we conclude that y^2 has the form $8n - 1$, where n is an integer. Then

$$\begin{aligned} x^2 - 3y^2 &= (2x_1)^2 - 3(8n + 1) = 4x_1^2 - 24n - 3 \\ &= 4(x_1^2 - 6n - 1) + 1 = 4l + 1, \end{aligned}$$

where $l = x_1^2 - 6n - 1$ is an integer. But an integer of the form $4l + 1$ cannot have the value -1; if it did, $4l = -2$, and therefore $l = -\frac{1}{2}$ would not be an integer. We leave it to the reader to show that if $x^2 - 3y^2 = -1$, then we cannot have x odd and y even.

Hence there do not exist integral solutions x, y of the equation

$$x^2 - 3y^2 = -1.$$

In fact, whenever N is such that $N - 3$ is an integral multiple of 4, the equation $x^2 - Ny^2 = -1$ has no solutions. On the other hand, if $N = p$ is a prime number of the form $4k + 1$, then the equation $x^2 - py^2 = -1$ always has solutions.

This last equation is closely connected with a famous theorem stated by Fermat in 1640 and proved by Euler in 1754:

THEOREM: *Every prime p of the form $4k + 1$ can be expressed as the sum of two squares, and this representation is unique. That is, there exists one and only one pair of integers P, Q such that $p = P^2 + Q^2$.*

Once this theorem became known it was natural for mathematicians to search for ways to calculate the numbers P and Q in terms of the given prime p. Constructions were given by Legendre (1808), Gauss (1825), Serret (1848), and others. Without entering the details of the proof, we shall present the essential idea of Legendre's construction.

Legendre's method depends upon the fact that the periodic part of the continued fraction for

$$\sqrt{p} = [a_1, \overline{a_2, a_3, \cdots, a_n, 2a_1}] = [a_1, \overline{a_2, a_3, a_4, \cdots, a_4, a_3, a_2, 2a_1}]$$

has a symmetrical part $a_2, a_3, a_4, \cdots, a_4, a_3, a_2$ followed by $2a_1$. We proved, however, in Section 4.8, that if the symmetrical part has no central term (n odd), then the equation $x^2 - py^2 = -1$ is soluble. The converse is also true, namely, *if $x^2 - py^2 = -1$ is soluble then there is no central term in the symmetrical part of the period;* hence the continued fraction for $\sqrt{p}$ has the form

$$\sqrt{p} = [a_1, \overline{a_2, a_3, \cdots, a_m, a_m, \cdots, a_3, a_2, 2a_1}].$$

This we write in the equivalent form

$$\sqrt{p} = a_1 + \frac{1}{a_2 +} \cdots \frac{1}{+ a_m +} \frac{1}{\alpha_{m+1}},$$

where, beginning at the middle of the symmetrical part,

$$\alpha_{m+1} = [\overline{a_m, a_{m-1}, \cdots, a_3, a_2, 2a_1, a_2, a_3, \cdots, a_m}].$$

Now α_{m+1} is a purely periodic continued fraction and hence has the form (see Theorem 4.1)

$$\alpha_{m+1} = \frac{P + \sqrt{p}}{Q}.$$

Moreover, the period in the expansion of α_{m+1} is symmetrical and hence the number β, obtained from α_{m+1} by reversing its period, is equal to α_{m+1}.

But according to Theorem 4.1, the conjugate α'_{m+1} of α_{m+1} is related to β by

$$\alpha'_{m+1} = -\frac{1}{\beta},$$

so that

$$\alpha'_{m+1} \cdot \beta = \alpha'_{m+1} \cdot \alpha_{m+1} = -1.$$

This means that

$$\frac{P+\sqrt{p}}{Q} \cdot \frac{P-\sqrt{p}}{Q} = -1 \qquad \text{or that} \qquad p = P^2 + Q^2.$$

As an illustration, take $p = 13 = 4 \cdot 3 + 1$. Expanding $\sqrt{13}$ we obtain

$$\sqrt{13} = [a_1, \overline{a_2, a_3, a_3, a_2, 2a_1}] = [3, \overline{1, 1, 1, 1, 6}],$$

so that

$$\alpha_{m+1} = \alpha_3 = [\overline{1, 1, 6, 1, 1}].$$

Hence all we have to do is calculate α_3. Thus

$$\sqrt{13} = 3 + \frac{1}{\alpha_1},$$

$$\alpha_1 = \frac{3+\sqrt{13}}{4} = 1 + \frac{1}{\alpha_2},$$

$$\alpha_2 = \frac{1+\sqrt{13}}{3} = 1 + \frac{1}{\alpha_3},$$

$$\alpha_3 = \frac{2+\sqrt{13}}{3} = \frac{P+\sqrt{p}}{Q},$$

so that $P = 2$, $Q = 3$, giving

$$p = 13 = 2^2 + 3^2.$$

Problem Set 22

1. Express $p = 29$ as the sum of two squares.
2. Express $p = 433$ as the sum of two squares.
3. There are two equal detachments of soldiers arranged in two squares, each containing b rows of b soldiers. Show that it is impossible to combine the two squares into a single square of soldiers.

   Show also that, if one soldier is added or taken away from one of the squares, the two detachments can sometimes be combined into a square

APPENDIX II

Some Miscellaneous Expansions

The following is a small collection of miscellaneous continued fractions, mainly of historical interest.† The list is not restricted to simple continued fractions.

1. Bombelli, 1572. In modern notation he knew essentially that

$$\sqrt{13} = 3 + \cfrac{4}{6 + \cfrac{4}{6 + \ddots}}\,.$$

2. Cataldi, 1613. He expressed the continued fraction expansion of $\sqrt{18}$ in the form

$$\sqrt{18} = 4 \cdot \& \frac{2}{8} \cdot \quad \& \frac{2}{8} \cdot \quad \& \frac{2}{8} \cdot \quad \ddots$$

and also in the form

$$\sqrt{18} = 4 \cdot \& \frac{2}{8.} \& \frac{2}{8.} \& \frac{2}{8.} \cdots.$$

† See D. E. Smith [13].

3. Lord Brouncker, about 1658.

$$\frac{4}{\pi} = 1 + \cfrac{1}{2 + \cfrac{9}{2 + \cfrac{25}{2 + \cfrac{49}{2 + \cfrac{81}{2 + \ddots}}}}}.$$

This expansion is closely connected historically with the infinite product

$$\frac{\pi}{2} = \frac{2 \cdot 2 \cdot 4 \cdot 4 \cdot 6 \cdot 6 \cdot 8 \cdot 8}{1 \cdot 3 \cdot 3 \cdot 5 \cdot 5 \cdot 7 \cdot 7 \cdot 9} \cdots$$

given by Wallis in 1655; both discoveries were important steps in the history of $\pi = 3.14159 \cdots$.

4. Euler, 1737. He found the following expansions involving the number

$$e = 2.7182818284590 \cdots = \lim_{n \to \infty} \left(1 + \frac{1}{n}\right)^n,$$

the base of the natural logarithms.

$$e - 1 = 1 + \cfrac{1}{1 + \cfrac{1}{2 + \cfrac{1}{1 + \cfrac{1}{1 + \cfrac{1}{4 + \ddots}}}}}$$

$$= [1, 1, 2, 1, 1, 4, 1, 1, 6, 1, 1, 8, \cdots].$$

$$\frac{e - 1}{e + 1} = \cfrac{1}{2 + \cfrac{1}{6 + \cfrac{1}{10 + \cfrac{1}{14 + \ddots}}}}.$$

$$\frac{e-1}{2} = \cfrac{1}{1+\cfrac{1}{6+\cfrac{1}{10+\cfrac{1}{14+\ddots}}}}.$$

This last expansion affords a quick approximation to e. For example, the 7th convergent to $(e-1)/2$ is $342762/398959$, so that, approximately,

$$e = \frac{1084483}{398959} = 2.718281828458\cdots.$$

This number differs from the value of e by one unit in the 12th decimal place.

5. Lambert, 1766.

$$\frac{e^x-1}{e^x+1} = \cfrac{1}{\cfrac{2}{x}+\cfrac{1}{\cfrac{6}{x}+\cfrac{1}{\cfrac{10}{x}+\cfrac{1}{\cfrac{14}{x}+\ddots}}}}.$$

$$\tan x = \cfrac{1}{\cfrac{1}{x}-\cfrac{1}{\cfrac{3}{x}-\cfrac{1}{\cfrac{5}{x}-\cfrac{1}{\cfrac{7}{x}-\ddots}}}}.$$

Lambert used these expansions to conclude that

a) If x is a rational number, not 0, then e^x cannot be rational;

b) If x is a rational number, not 0, then $\tan x$ cannot be rational.

Thus, since $\tan(\pi/4) = 1$, neither $\pi/4$ nor π can be rational.

Some weaknesses in Lambert's proof were remedied by Legendre in his *Éléments de géometrie* (1794).

6. Lambert, 1770.

$$\pi = 3+\cfrac{1}{7+\cfrac{1}{15+\cfrac{1}{1+\cfrac{1}{292+\ddots}}}}$$

$$= [3, 7, 15, 1, 292, 1, 1, 1, 2, 1, 3, 1, 14, 2, 1, 1, 2, 2, 2, 2, 1, 84, 2, \cdots].$$

Unlike the expansion of e, the simple continued fraction expansion of $\pi = 3.1415926536 \cdots$ does not seem to have any regularity. The convergents to π are

$$\frac{3}{1}, \frac{22}{7}, \frac{333}{106}, \frac{355}{113}, \frac{103993}{33102}, \frac{104348}{33215}, \cdots;$$

the fraction

$$\frac{355}{113} = 3.14159292035 \cdots$$

approximates π with an error of at most 3 units in the 7th decimal place.

7.

$$\sqrt{a^2 + b} = a + \cfrac{b}{2a + \cfrac{b}{2a + \cfrac{b}{2a + \cdots}}}, \qquad a^2 + b > 0.$$

8.

$$\sqrt{2} = 1 + \cfrac{1}{2 + \cfrac{1}{2 + \cfrac{1}{2 + \cdots}}}.$$

9.

$$\frac{1 + \sqrt{5}}{2} = 1 + \cfrac{1}{1 + \cfrac{1}{1 + \cfrac{1}{1 + \cdots}}}.$$

The convergents are $\frac{1}{1}, \frac{2}{1}, \frac{3}{2}, \frac{5}{3}, \frac{8}{5}, \cdots$, both numerators and denominators being formed from the sequence of Fibonacci numbers 1, 1, 2, 3, 5, 8, 13, $\cdots$.

10. Stern, 1833.

$$\frac{\pi}{2} = 1 - \cfrac{1}{3 - \cfrac{2 \cdot 3}{1 - \cfrac{1 \cdot 2}{3 - \cfrac{4 \cdot 5}{1 - \cfrac{3 \cdot 4}{3 - \cfrac{6 \cdot 7}{1 - \cfrac{5 \cdot 6}{3 - \cdots}}}}}}}$$

11.

$$\sin x = \cfrac{x}{1 + \cfrac{x^2}{(2 \cdot 3 - x^2) + \cfrac{2 \cdot 3x^2}{(4 \cdot 5 - x^2) + \cfrac{4 \cdot 5x^2}{(6 \cdot 7 - x^2) + \ddots}}}}.$$

12. Lambert, 1770.

$$\tan x = \cfrac{x}{1 - \cfrac{x^2}{3 - \cfrac{x^2}{5 - \cfrac{x^2}{7 - \ddots}}}}.$$

13. Gauss, 1812.

$$\tanh x = \cfrac{x}{1 + \cfrac{x^2}{3 + \cfrac{x^2}{5 + \ddots}}}.$$

14. Lambert, 1770; Lagrange, 1776.

$$\arctan x = \cfrac{x}{1 + \cfrac{1 \cdot x^2}{3 + \cfrac{4 \cdot x^2}{5 + \cfrac{9x^2}{7 + \cfrac{16x^2}{9 + \ddots}}}}}, \qquad |x| < 1.$$

15. Lambert, 1770; Lagrange, 1776.

$$\log(1 + x) = \cfrac{x}{1 + \cfrac{1^2x}{2 + \cfrac{1^2x}{3 + \cfrac{2^2x}{4 + \cfrac{2^2x}{5 + \cfrac{3^2x}{6 + \cfrac{3^2x}{7 + \ddots}}}}}}}, \qquad |x| < 1.$$

16. Lagrange, 1813.

$$\log\frac{1+x}{1-x} = \cfrac{2x}{1-\cfrac{1\cdot x^2}{3-\cfrac{4x^2}{5-\cfrac{9x^2}{7-\cfrac{16x^2}{9-\ddots}}}}}, \qquad |x|<1.$$

17. Lagrange, 1776.

$$(1+x)^k = \cfrac{1}{1-\cfrac{kx}{1+\cfrac{\frac{1\cdot(1+k)}{1\cdot 2}x}{1+\cfrac{\frac{1\cdot(1-k)}{2\cdot 3}x}{1+\cfrac{\frac{2(2+k)}{3\cdot 4}x}{1+\cfrac{\frac{2(2-k)}{4\cdot 5}x}{1+\cfrac{\frac{3(3+k)}{5\cdot 6}x}{1+\ddots}}}}}}}, \qquad |x|<1.$$

18. Laplace, 1805; Legendre, 1826.

$$\int_0^x e^{-u^2}\,du = \frac{\sqrt{\pi}}{2} - \cfrac{\frac{1}{2}e^{-x^2}}{x+\cfrac{1}{2x+\cfrac{2}{x+\cfrac{3}{2x+\cfrac{4}{x+\ddots}}}}}, \qquad x>0.$$

This is the probability integral used in the theory of probability and in statistics.

Solutions

Set 1, page 13

1. (a) [1, 1, 1, 5] (b) [1, 1, 1, 5] (c) [3, 1, 1, 5, 1, 3]
   (d) [1, 3, 6, 4, 2] (e) [0, 4, 2, 1, 7] (f) [3, 2, 1, 6, 2, 2]
   (g) [3, 7, 15, 1, 25, 1, 7, 4]

2. $\frac{93}{29}$.

3. $\frac{11}{31}$.

4. $\frac{355}{113} = 3.1415929204 \cdots$, and $\pi = 3.1415926536 \cdots$.

5. (a) [0, 1, 1, 1, 5] (b) [0, 1, 1, 1, 5]

6. If $p > q > 0$, then

$$\frac{p}{q} > 1 \text{ and } \frac{p}{q} = [a_1, a_2, \cdots, a_n] = a_1 + \cfrac{1}{a_2 + \cfrac{1}{a_3 + \ddots + \cfrac{1}{a_n}}},$$

where a_1 is an integer > 0. The reciprocal of $\frac{p}{q}$ is

$$\frac{q}{p} = \frac{1}{\frac{p}{q}} = \cfrac{1}{a_1 + \cfrac{1}{a_2 + \cfrac{1}{a_3 + \ddots + \cfrac{1}{a_n}}}}$$

$$= 0 + \cfrac{1}{a_1 + \cfrac{1}{a_2 + \ddots + \cfrac{1}{a_n}}}$$

$$= [0, a_1, a_2, \cdots, a_n].$$

Conversely, if $q < p$, then $\dfrac{q}{p}$ is of the form

$$\frac{q}{p} = 0 + \cfrac{1}{a_1 + \cfrac{1}{a_2 + \ddots + \cfrac{1}{a_n}}}$$

and its reciprocal is

$$\frac{p}{q} = \cfrac{1}{\cfrac{1}{a_1 + \cfrac{1}{a_2 + \ddots + \cfrac{1}{a_n}}}} = a_1 + \cfrac{1}{a_2 + \cfrac{1}{a_3 + \ddots + \cfrac{1}{a_n}}}.$$

Set 2, page 19

1. (a) $[5, 1, 3, 1]$, $[5, 1, 4]$, (b) $[0, 5, 1, 4]$, $[0, 5, 1, 3, 1]$,
(c) $[-6, 5]$, $[-6, 4, 1]$, (d) $[3, 1, 29, 1]$, $[3, 1, 30]$,
(e) $[-4, 31]$, $[-4, 30, 1]$, (f) $[0, 3, 1, 30]$, $[0, 3, 1, 29, 1]$.

2. (a) 69, (b) 1, (c) 19, (d) 21.

Set 3, page 25

1. (a) $[5, 1, 3, 5]$, convergents $\frac{5}{1}, \frac{6}{1}, \frac{23}{4}, \frac{121}{21}$.
(b) $[3, 1, 1, 2, 1, 1, 1, 1, 2]$, convergents $\frac{3}{1}, \frac{4}{1}, \frac{7}{2}, \frac{18}{5}, \frac{25}{7}, \frac{43}{12}, \frac{68}{19}, \frac{111}{31}, \frac{290}{81}$.
(c) $[0, 1, 1, 1, 1, 5, 1, 8]$, convergents $\frac{0}{1}, \frac{1}{1}, \frac{1}{2}, \frac{2}{3}, \frac{3}{5}, \frac{17}{28}, \frac{20}{33}, \frac{177}{292}$.
(d) $[5, 2, 11]$, convergents $\frac{5}{1}, \frac{11}{2}, \frac{126}{23}$.

2. (a) $[2, 1, 1, 4, 2]$, (b) $[4, 2, 1, 7, 8]$,
(c) $[0, 4, 2, 5, 1]$, (d) $[4, 2, 7]$.

3. Even number of quotients: (a) $p_6/q_6 = \frac{51}{20}$, $p_5/q_5 = \frac{28}{11}$, hence $p_6q_5 - p_5q_6 = 51 \cdot 11 - 28 \cdot 20 = 561 - 560 = 1$, (b) 1, (c) 1, (d) 1.

Odd number of quotients: (a) $[2, 1, 1, 4, 2]$, $p_5/q_5 = \frac{51}{20}$, $p_4/q_4 = \frac{23}{9}$, hence $p_5q_4 - p_4q_5 = 51 \cdot 9 - 23 \cdot 20 = 459 - 460 = -1$, (b) -1, (c) -1, (d) -1.

4. $1393 = 5 \cdot 225 + 5 \cdot 43 + 4 \cdot 10 + 3 \cdot 3 + 2 \cdot 1 + 2$.

5. $p_5/p_4 = \frac{134}{23} = [5, 1, 4, 1, 3]$; compare with original fraction. Similarly, $q_5/q_4 = \frac{35}{6} = [5, 1, 5] = [5, 1, 4, 1]$.

6. (a) $\frac{3}{1}, \frac{22}{7}, \frac{333}{106}, \frac{355}{113}, \frac{9208}{2931}, \frac{9563}{3044}, \frac{76149}{24239}, \frac{314159}{100000}.$

(b) $\frac{2}{1}, \frac{3}{1}, \frac{8}{3}, \frac{11}{4}, \frac{19}{7}, \frac{87}{32}, \frac{106}{39}, \frac{1359}{500} = \frac{2718}{1000}.$

(c) $\frac{0}{1}, \frac{1}{2}, \frac{10}{21}, \frac{21}{44}, \frac{52}{109}, \frac{73}{153}, \frac{125}{262}, \frac{2323}{4869}, \frac{4771}{10000}.$

(d) $\frac{0}{1}, \frac{1}{3}, \frac{3}{10}, \frac{28}{93}, \frac{31}{103}, \frac{90}{299}, \frac{301}{1000}.$

7. From $p_n = a_n p_{n-1} + p_{n-2}$ we see that

$$\frac{p_n}{p_{n-1}} = a_n + \cfrac{1}{\cfrac{p_{n-1}}{p_{n-2}}},$$

and from the fact that $p_{n-1} = a_{n-1}p_{n-2} + p_{n-3}$ we see that

$$\frac{p_{n-1}}{p_{n-2}} = a_{n-1} + \cfrac{1}{\cfrac{p_{n-2}}{p_{n-3}}}.$$

Similarly

$$\frac{p_{n-2}}{p_{n-3}} = a_{n-2} + \cfrac{1}{\cfrac{p_{n-3}}{p_{n-4}}}$$

....

$$\frac{p_3}{p_2} = a_3 + \cfrac{1}{\cfrac{p_2}{p_1}} = a_3 + \frac{1}{a_2 +} \frac{1}{a_1}.$$

The required result is then obtained from these equations by successive substitutions. The result for q_n/q_{n-1} is proved in a like manner.

8. In constructing our table of convergents, we used the fact that $p_n = np_{n-1} + p_{n-2}$. In this relation let n have in turn the value $n, n-1, n-2, \cdots, 3, 2, 1$. This gives the following equations:

$$p_n = np_{n-1} + p_{n-2}$$

$$p_{n-1} = (n-1)p_{n-2} + p_{n-3}$$

$$p_{n-2} = (n-2)p_{n-3} + p_{n-4}$$

....

$$p_3 = 3p_2 + p_1$$

$$p_2 = 2p_1 + 1$$

Adding the left and right sides of these equations, we obtain

$$\begin{aligned} p_n + p_{n-1} + \quad & p_{n-2} + \quad \cdots \quad + p_3 + p_2 + p_1 \\ = np_{n-1} + \quad & p_{n-2} \\ & + (n-1)p_{n-2} + p_{n-3} \\ & \qquad + (n-2)p_{n-3} + p_{n-4} \\ & \qquad \cdots\cdots\cdots\cdots \\ & \qquad\qquad + 3p_2 + p_1 \\ & \qquad\qquad\quad + 2p_1 + 1 \\ & \qquad\qquad\quad + p_1. \end{aligned}$$

Leaving p_n on the left and subtracting the terms $p_{n-1}, p_{n-2}, \cdots p_2, p_1$ from both sides of the equation, we obtain the required expression, namely

$$p_n = (n-1)p_{n-1} + (n-1)p_{n-2} + (n-2)p_{n-3} + \cdots + 3p_2 + 2p_1 + (p_1 + 1).$$

Set 4, page 28

1. $p_0q_{-1} - p_{-1}q_0 = 1 \cdot 1 - 0 \cdot 0 = 1 = (-1)^0$, $p_1q_0 - p_0q_1 = 3 \cdot 0 - 1 \cdot 1 = (-1)^1$, $p_2q_1 - p_1q_2 = 4 \cdot 1 - 3 \cdot 1 = (-1)^2$, etc. The second part of the problem is accomplished by simple calculations.

Set 5, page 35

1. (a) Show that $x = -3y + t$ when $t = \dfrac{2-2y}{15}\cdot$ Thus, $y = 1 - 7t - u$, where $u = t/2$ or $t = 2u$. Hence

$$y = 1 - 7(2u) - u = 1 - 15u,$$
$$x = -3(1 - 15u) + 2u = -3 + 47u.$$

Required solution is

$$x = -3 + 47u, \quad y = 1 - 15u, \quad u = 0, \pm 1, \pm 2, \pm 3, \cdots.$$

For both x and y to be positive integers, u must be an integer such that $u < \frac{1}{15}$ and $u > \frac{3}{47}$. Clearly, no such integer exists; hence, there are no integral solutions with both x and y positive. *Note* that some solutions might have a different parametric form but still reproduce the same values of x and y.

(b) $x = -2 + 7u$, $y = 9 - 31u$, $u = 0, \pm 1, \pm 2, \cdots$. There are no positive integral solutions; for, no integer u can be simultaneously less than $\frac{9}{31}$ and greater than $\frac{2}{7}$.

(c) $x = -6 + 47u, \quad y = 2 - 15u, \quad u = 0, \pm 1, \pm 2, \cdots$. No positive solutions.

(d) $x = 34 - 21w, \quad y = 13w - 7, \quad w = 0, \pm 1, \pm 2, \cdots$. For positive solutions w must be an integer less than $\frac{34}{21}$ and greater than $\frac{7}{13}$. Hence $w = 1$, and the only positive solution is $(x, y) = (13, 6)$.

2. The given equation has no integral solutions. By Euler's method you would arrive at other equations which cannot be solved in integers. For example $x = 3 - 3y - u$ where $u = \frac{1}{6}(1 + 3y)$. But no integral value of y will make u an integer. Why?

3. The straight line if carefully graphed should pass through the two points $(x, y) = (2, 13)$ and $(x, y) = (7, 5)$.

4. Let x = number of horses and let y = number of cows. Then $37x + 22y = 2370$. The general solution is $x = 22t + 4$, $y = 101 - 37t$. For positive solutions t must be an integer between $-\frac{2}{11}$ and $\frac{101}{37}$. Hence $t = 0, 1, 2$, and the positive solutions are $(x, y) = (4, 101), (26, 64), (48, 27)$.

5. $x = 15u - 5$, $y = 17u - 6$. Positive solutions require that $u > \frac{1}{3}$ and $u > \frac{6}{17}$, hence $u = 1, 2, 3, \cdots$.

6. The solution of the equation $9x + 13y = u + v = 84$ is $x = 5 + 13t$, $y = 3 - 9t$. Hence $u = 9(5 + 13t)$, $v = 13(3 - 9t)$, where t is any integer.

7. The solution of the equation $2x - 3y = 1$ is $x = 3u - 1$, $y = 2u - 1$. Hence $N = 20x + 2 = 60u - 18$, where u is any integer. For example, when $u = -1$,

$$N = -78 = -4(20) + 2 = -3(30) - 12.$$

Set 6, page 42

(In all cases $t = 0, \pm 1, \pm 2, \pm 3, \cdots$)

1. (a) $\frac{13}{17} = [0, 1, 3, 4]$, $\quad n = 4$, $\quad x_0 = q_3 = 4$, $\quad y_0 = p_3 = 3$, hence

$$x = x_0 + tb = 4 + 17t, \qquad y = y_0 + ta = 3 + 13t.$$

(b) $\frac{13}{17} = [0, 1, 3, 3, 1]$, $\quad n = 5$, $\quad x_0 = q_4 = 13$, $\quad y_0 = p_4 = 10$, hence

$$x = x_0 + tb = 13 + 17t, \qquad y = y_0 + ta = 10 + 13t.$$

(c) $\frac{65}{56} = [1, 6, 4, 2]$, $\quad n = 4$, $\quad x_0 = q_3 = 25$, $\quad y_0 = p_3 = 29$, hence

$$x = 25 + 56t, \qquad y = 29 + 65t.$$

(d) $\frac{65}{56} = [1, 6, 4, 1, 1]$, $\quad n = 5$, $\quad x_0 = q_4 = 31$, $\quad y_0 = p_4 = 36$, hence

$$x = 31 + 56t, \qquad y = 36 + 65t.$$

(e) $\frac{56}{65} = [0, 1, 6, 4, 1, 1]$, $\quad n = 6$, $\quad x_0 = q_5 = 36$, $\quad y_0 = p_5 = 31$, hence

$$x = 36 + 65t, \qquad y = 31 + 56t.$$

Set 7, page 44

(In all cases $t = 0, \pm 1, \pm 2, \cdots$)

1. (a) $x_0 = 4, \ y_0 = 3, \ c = 5, \ x = cx_0 + bt = 20 + 17t,$
$y = cy_0 + at = 15 + 13t$. Check: $13(20 + 17t) - 17(15 + 13t) = 5$.
(b) $x_0 = 25, \ y_0 = 29, \ c = 7, \ x = cx_0 + bt = 175 + 56t,$
$y = cy_0 + at = 203 + 65t.$
(c) $x_0 = 29, \ y_0 = 25,$ is a particular solution of $56x - 65y = -1$; hence $c = 3, \ x = cx_0 + bt = 87 + 65t, \ y = cy_0 + at = 75 + 56t.$

Set 8, page 48

1. (a) The g.c.d. of $183(= 3 \cdot 61)$ and $174(= 2 \cdot 3 \cdot 29)$ is 3, and since 3 divides 9 the equation is solvable. Divide both sides of the given equation by 3 and solve the resulting equation $61x + 58y = 3$. We first solve the equation $61x - 58y = 1$ for which the expansion $\frac{61}{58} = [1, 19, 2, 1]$ shows that $x_0 = q_{n-1} = 39, \ y_0 = p_{n-1} = 41$. Hence the solution of the given equation, according to equation (2.28), is

$$x = cq_{n-1} - tb = 3 \cdot 39 - 58t = 117 - 58t,$$
$$y = at - cp_{n-1} = 61t - 3 \cdot 41 = 61t - 123.$$

(b) In this case we must solve the equation $61x - 58y = 3$, from which the solution of the given equation according to (2.23) is

$$x = cx_0 + bt = 117 + 58t, \qquad y = cy_0 + at = 123 + 61t.$$

(c) An unsolvable equation since $77 = 7 \cdot 11$ and $63 = 3^2 \cdot 7$ so the g.c.d. of 77 and 63 is 7 and does not divide $40(= 2^3 \cdot 5)$.
(d) Since $34(= 2 \cdot 17)$ and $49(= 7^2)$ are relatively prime, we have only to solve the given equation by the methods of Section 2.4. The required solution is $x = 65 + 49t, \ y = 45 + 34t$.
(e) $x = 65 - 49t, \quad y = 34t - 45.$
(f) The g.c.d. of $56(= 2^3 \cdot 7)$ and $20(= 2^2 \cdot 5)$ is 4 and does not divide 11. The given equation has no integral solutions.

2. The solution of the equation $11x + 7y = 68$ is $x = 136 - 7t$, $y = 11t - 204$. The only solution with both x and y positive is $x = 3, \ y = 5$ given by $t = 19$.

3. From the hint we find that $7x + 9y = 90$. The general solution of this equation is $x = 360 - 9t, \ y = 7t - 270$. For positive values of a and b it is sufficient to require that $x \geq 0, \ y \geq 0$, or that t be an integer $\leq 360/9$ and $> 270/7$. Thus we can take $t = 39$ and $t = 40$.
When $t = 39, \ x = 9, \ y = 3$ and $a = 68, b = 32$.
When $t = 40, \ x = 0, \ y = 10$ and $a = 5, \ b = 95$.

4. The general solution is $x = 1200 - 17t, \ y = 13t - 900$. The value $t = 70$ leads to the only positive solution, $x = 10, \ y = 10$.

Set 9, page 59

1. Expansions are given in the problems. The first five convergents are:

(a) $\frac{2}{1}, \frac{5}{2}, \frac{22}{9}, \frac{49}{20}, \frac{218}{89}$ (c) $\frac{6}{1}, \frac{7}{1}, \frac{13}{2}, \frac{46}{7}, \frac{59}{9}$

(b) $\frac{2}{1}, \frac{3}{1}, \frac{5}{2}, \frac{8}{3}, \frac{37}{14}$ (d) $\frac{1}{1}, \frac{6}{5}, \frac{13}{11}, \frac{45}{38}, \frac{103}{87}$

(e) $\frac{0}{1}, \frac{1}{3}, \frac{1}{4}, \frac{3}{11}, \frac{4}{15}$

2. (a) Let $x = [2, \overline{2, 4}] = 2 + \frac{1}{y}$, $y = 2 + \frac{1}{4+}\frac{1}{y}$. Hence

$$y = \frac{2 + \sqrt{6}}{2} = \frac{1}{\sqrt{6} - 2} \quad \text{and} \quad x = 2 + (\sqrt{6} - 2) = \sqrt{6}.$$

(b) Let $x = 5 + \frac{1}{y}$, $y = 1 + \frac{1}{1+}\frac{1}{1+}\frac{1}{10+}\frac{1}{y}$. Then

$$7y^2 - 10y - 1 = 0, \quad \text{or} \quad y = \frac{5 + \sqrt{32}}{7} = \frac{1}{\sqrt{32} - 5}.$$

Hence $x = 5 + (\sqrt{32} - 5) = \sqrt{32}$.

3. To show that $AH = \frac{4^2}{7^2 + 8^2}$, see Figure 3, note that in triangle AOD, $(AD)^2 = \frac{7^2 + 8^2}{8^2}$. From the similar triangles AGF and AOD we see that

$$\frac{AF}{AG} = \frac{AD}{1} \quad \text{or} \quad \frac{(AF)^2}{(AG)^2} = AD^2 \quad \text{or} \quad (AG)^2 = \frac{(AF)^2}{(AD)^2},$$

and, since $AF = \frac{1}{2}$,

$$(AG)^2 = \frac{4^2}{7^2 + 8^2}.$$

On the other hand from the similar triangles AHF and AGD we see that $AF/AD = AH/AG$. But we know already that $AF/AD = AG/1$. Hence, by division,

$$1 = \frac{AH}{(AG)^2} \quad \text{or} \quad AH = \frac{4^2}{7^2 + 8^2}.$$

6. In the nth year, the total number of branches, F_n, consists of the number of branches O_n that are at least one year old and the number of branches

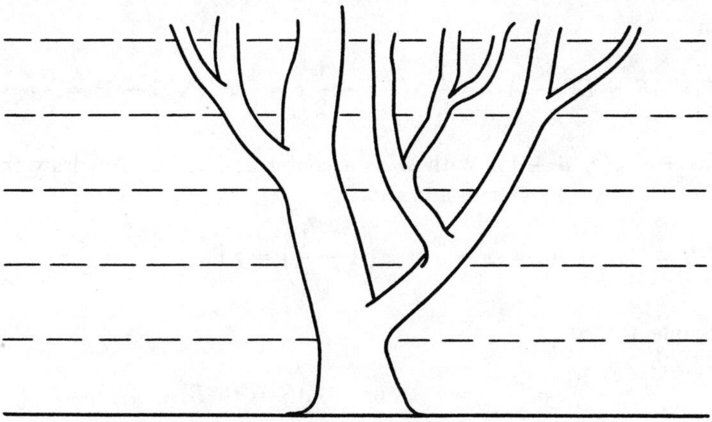

Figure 12

Y_n that are less than one year old. In symbols, $F_n = O_n + Y_n$. During the next year, there are

$$F_{n+1} = 2O_n + Y_n = O_n + O_n + Y_n = F_n + O_n$$

branches. Since the number of at-least-one-year-old branches constitutes the total number of branches of the previous year, $O_n = F_{n-1}$. Thus

$$F_{n+1} = F_n + F_{n-1} \qquad \text{for} \qquad n = 2, 3, \cdots$$

and $F_1 = 1$ (because only the trunk was present during the first year) yield the recursion formula for these Fibonacci numbers.

8. *First solution:* Construct the square $ABCD$ of side $x = AB$; see Figure 13a. Construct the point E such that $AE = ED$ and draw $EB = \frac{1}{2}\sqrt{5}x$. With E as center and radius EB describe the arc BF.

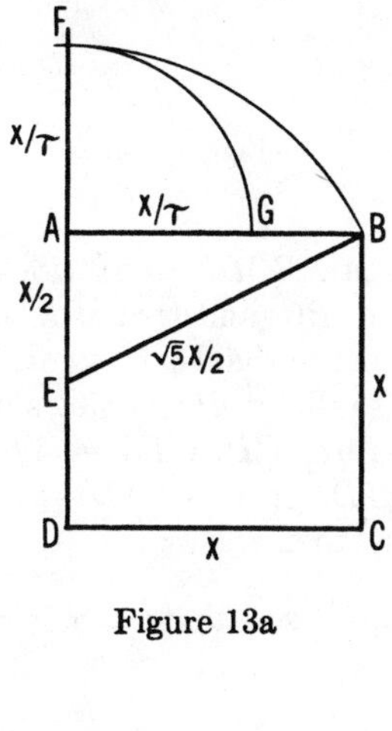

Figure 13a

Then

$$AG = AF = EF - AE = \frac{1}{2}\sqrt{5}\,x - \frac{1}{2}x = \frac{1}{2}x\cdot(\sqrt{5}-1) = \frac{x}{\tau},$$

where $\tau = \frac{1}{2}(\sqrt{5}+1)$. With A as a center and radius AF draw the arc FG. Clearly $AG = x/\tau$ and so

$$GB = AB - AG = x - \frac{x}{\tau} = x\left(1 - \frac{1}{\tau}\right) = x\left(\frac{\tau - 1}{\tau}\right) = \frac{x}{\tau^2}.$$

Consequently,

$$\frac{AG}{GB} = \frac{x/\tau}{x/\tau^2} = \tau \qquad \text{or} \qquad AG = \tau(GB).$$

Second solution: Construct the right triangle BAC such that $AB = x$, $AC = \frac{1}{2}x$; see Figure 13b. With C as center and radius $BC = \frac{1}{2}\sqrt{5}\,x$ construct point D. Then $AC + CD = \tau x$, where $\tau = \frac{1}{2}(1+\sqrt{5})$. Construct point E such that $DE = AB = x$. Draw BE, and GD parallel to BE. Then

$$\frac{AG}{GB} = \frac{AD}{DE} = \frac{\tau x}{x} = \tau, \qquad \text{and} \qquad AG = \tau(GB).$$

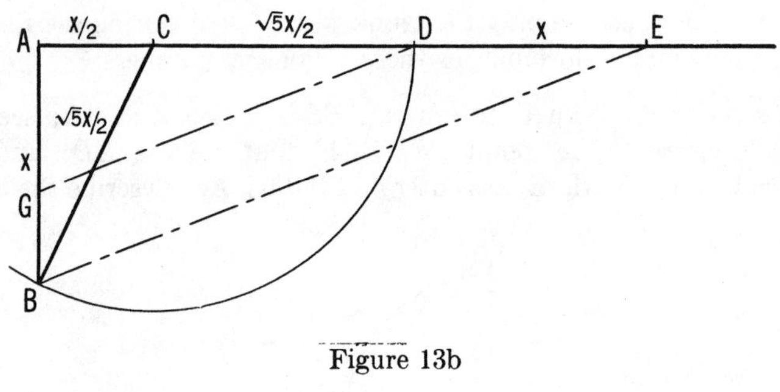

Figure 13b

9. For the regular pentagon $ABCDE$ whose sides have length 1, first prove that AD is parallel to BC and that BE is parallel to CD. Hence $BG = CD$. Similarly, prove that HI is parallel to BE, and that BH is parallel to FI, and so $BF = HI$. Using similar triangles, we see that $AD/AI = CD/HI$. But $CD = BG = AI$, and $HI = BF = ID$, hence $AD/AI = AI/ID$, or $(AD)(ID) = (AI)^2$. Now $BC = 1 = AI$, and if we let $AD = x$, then $ID = x - 1$, and $x(x-1) = 1$, or

$$x^2 - x - 1 = 0, \qquad \text{so that} \qquad x = \tau = \tfrac{1}{2}(1+\sqrt{5}).$$

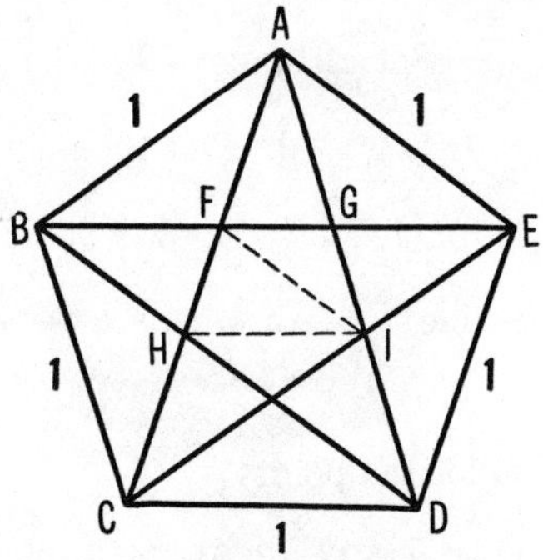

Figure 14

A line segment of length τ can be constructed using the results of Problem 4. Hence to construct a regular pentagon, draw line $CD = 1$; and with C and D as centers and radius $AC = CD = \tau$, construct point A. Points B and E can then be constructed since $AB = BC = AE = DE = 1$.

Set 10, page 63

1. The odd convergents to $\sqrt{2} = [1, \bar{2}]$ are $\frac{1}{1}, \frac{7}{5}\ \frac{41}{29}, \cdots$ and all are less than $\sqrt{2} = 1.414 \cdots$. The even convergents are $\frac{3}{2}, \frac{17}{12}, \frac{99}{70}, \cdots$ and are all greater than $\sqrt{2}$. Moreover $\frac{1}{1} < \frac{7}{5} < \frac{3}{2}$, etc.

Set 11, page 76

1. $\frac{2893}{1323} = [2, 5, 2, 1, 4, 5, 1, 2]$ and $c_1 = \frac{2}{1}$, $c_2 = \frac{11}{5}$, $c_1 = \frac{24}{11}$, $c_4 = \frac{35}{16}$, $c_5 = \frac{164}{75}$, $c_6 = \frac{855}{391}$, $\cdots$. A calculation shows that

$$\left|\frac{2893}{1323} - \frac{164}{75}\right| < \frac{1}{q_5 q_6} < 0.0005,$$

hence the required approximation is $\frac{164}{75}$. In this problem it would have sufficed to use $1/q_5^2$ in place of $1/q_5 q_6$ since $(\frac{1}{75})^2 < 0.0005$.

2. $\sqrt{19} = 4.358899 \cdots = [4, \overline{2, 1, 3, 1, 2, 8}]$. The convergents are $\frac{4}{1}$, $\frac{9}{2}$, $\frac{13}{3}$, $\frac{48}{11}$, $\frac{61}{14}$, $\frac{170}{39}$, $\frac{1421}{326}$, $\cdots$. The convergent $c_7 = \frac{1421}{326}$ gives $1/q_7^2 = 1/326^2 < 0.00005$, hence c_7 is the required approximation.

3. The first five convergents to π are $\frac{3}{1}$, $\frac{22}{7}$, $\frac{333}{106}$, $\frac{355}{113}$, $\frac{103993}{33102}$. Calculate, in turn,

$$\frac{1}{1 \cdot 7}, \quad \frac{1}{7 \cdot 106}, \quad \frac{1}{113 \cdot 33102}.$$

For example, $\dfrac{1}{7 \cdot 106} = 0.00134$; hence the error in using $\frac{22}{7}$ in place of π is at most $0.00134 \cdots$.

Set 12, page 79

1. $\alpha = \frac{1}{2}(\sqrt{5}-1) = [0, 1, 1, 1, \cdots]$. Plot the points

$$(x, y) = (q_n, p_n) = (1, 0), (1, 1), (2, 1), (3, 2), (5, 3), (8, 5), (13, 8), \cdots$$

Also carefully plot the line $y = \alpha x$ where $\alpha = 0.62$, the approximate value of $\frac{1}{2}(\sqrt{5}-1)$.

2. $\alpha = \sqrt{3} = [1, \overline{1, 2}]$. Plot the points

$$(x, y) = (q_n, p_n) = (1, 1),\ (1, 2),\ (3, 5),\ (4, 7),\ (11, 19),\ (15, 26), \cdots.$$

Also plot the line $y = \sqrt{3}\,x$, $\sqrt{3} = 1.732\cdots$.

Set 13, page 80

1. (a) $x = \frac{1}{2}(3+\sqrt{13}) = 3.30277$ · The values of the first few convergents are:

$$\frac{3}{1} = 3.0000\cdots,\quad \frac{10}{3} = 3.3333\cdots,\quad \frac{33}{10} = 3.3000\cdots,\quad \frac{109}{33} = 3.0303\cdots.$$

(b) $x = \frac{1}{2}(5+\sqrt{29}) = 5.19258\cdots$. The values of the first few convergents are:

$$\frac{5}{1} = 5.0000\cdots,\quad \frac{26}{5} = 5.2000\cdots,\quad \frac{135}{26} = 5.1923\cdots,\quad \frac{701}{135} = 5.1926\cdots.$$

2. Write

$$x = b + \cfrac{1}{a + \cfrac{1}{x}} = \frac{(ab+1)x + b}{ax+1};$$

then $ax^2 - abx - ac = 0$ since $b = ac$, and x also satisfies the equivalent equation $x^2 - bx - c = 0$.

3. For example, let $a = 1$, $b = 2$, then $x = [0, 1, 2, 1, 2, \cdots] = [0, \overline{1, 2}]$. The first few convergents are $\frac{0}{1}, \frac{1}{1}, \frac{2}{3}, \frac{3}{4}, \frac{8}{11}, \cdots$. Let

$$\frac{p_{n-2}}{q_{n-2}} = \frac{p_1}{q_1} = \frac{0}{1},\qquad \frac{p_n}{q_n} = \frac{p_3}{q_3} = \frac{2}{3},\qquad \frac{p_{n+2}}{q_{n+2}} = \frac{p_5}{q_5} = \frac{8}{11},$$

then $p_{n+2} - (ab+2)p_n + p_{n-2} = 0$ gives $8 - 4(2) + 0 = 0$. Try other cases.

Set 14, page 96

1. (a) $\alpha = \frac{1}{3}(\sqrt{12} + 3) > 1, \qquad \beta = \sqrt{12} + 3 > 1.$
   (b) $3\alpha^2 - 6\alpha - 1 = 0.$
   (c) $\alpha' = \frac{1}{3}(3 - \sqrt{12}) = -0.154 \cdots,$
   $-1/\beta = -1/(\sqrt{12} + 3) = \frac{1}{3}(3 - \sqrt{12}).$

2. (a) $2\alpha^2 + 2\alpha - 7 = 0, \qquad \alpha = \frac{1}{2}(\sqrt{15} - 1) > 1,$
   $\alpha' = \frac{1}{2}(-1 - \sqrt{15}) < -2.$
   (b) $3\gamma^2 - 5\gamma + 1 = 0, \qquad \gamma = \frac{1}{6}(5 + \sqrt{13}) > 1,$
   $\gamma' = \frac{1}{6}(5 - \sqrt{13}) = 0.232 \cdots > 0.$

Set 15, page 100

1. $\alpha_1 \pm \alpha_2 = (A_1 \pm A_2) + (B_1 \pm B_2)\sqrt{D};$

$$\alpha_1 \cdot \alpha_2 = (A_1 + B_1\sqrt{D}) \cdot (A_2 + B_2\sqrt{D})$$
$$= A_1A_2 + B_1B_2D + (A_1B_2 + A_2B_1)\sqrt{D}$$
$$\frac{\alpha_1}{\alpha_2} = \left(\frac{A_1A_2 - B_1B_2D}{A_2^2 - B_2^2D}\right) + \left(\frac{A_2B_1 - A_1B_2}{A_2^2 - B_2^2D}\right)\sqrt{D}, \qquad A_2^2 - B_2^2\,D \neq 0,$$

for if $A_2^2 - B_2^2D = 0$, then D would be a perfect square.

2. $(\alpha_1 - \alpha_2)' = (A_1 - A_2) - (B_1 - B_2)\sqrt{D}$

$$= (A_1 - B_1\sqrt{D}) - (A_2 - B_2\sqrt{D}) = \alpha_1' - \alpha_2';$$
$$(\alpha_1 \cdot \alpha_2)' = (A_1A_2 + B_1B_2D) - (A_1B_2 + B_1A_2)\sqrt{D}$$
$$= (A_1 - B_1\sqrt{D})(A_2 - B_2\sqrt{D}) = \alpha_1' \cdot \alpha_2';$$
$$\left(\frac{\alpha}{\alpha_2}\right)' = \left(\frac{A_1A_2 - B_1B_2D}{A_2^2 - B_2^2D}\right) - \left(\frac{A_2B_1 - A_1B_2}{A_2^2 - B_2^2D}\right)\sqrt{D};$$

on the other hand

$$\frac{\alpha_1'}{\alpha_2'} = \frac{A_1 - B_1\sqrt{D}}{A_2 - B_2\sqrt{D}} \cdot \frac{A_2 + B_2\sqrt{D}}{A_2 + B_2\sqrt{D}}$$
$$= \left(\frac{A_1A_2 - B_1B_2D}{A_2^2 - B_2^2D}\right) - \left(\frac{A_2B_1 - A_1B_2}{A_2^2 - B_2^2D}\right)\sqrt{D}.$$

3. $A + B\sqrt{M} = -C\sqrt{N}$; therefore $2AB\sqrt{M} = C^2N - A^2 - B^2M$. If $AB \neq 0$, the left side of this equation is irrational, the right side rational; this is impossible. If $AB = 0$, then $A = 0$ or $B = 0$. If $A = 0$, $B \neq 0$, then from $A + B\sqrt{M} + C\sqrt{N} = 0$ we see that $\sqrt{M}/\sqrt{N} = -C/B$, contrary to hypothesis. Hence if $A = 0$, then $B = 0$, and hence $C = 0$. If $B = 0$, $A + C\sqrt{N} = 0$, hence $A = 0$, $C = 0$.

Set 16, page 104

1. The largest integer less than $\frac{1}{3}(5+\sqrt{37})$ is 3. If

$$\alpha = \frac{5+\sqrt{37}}{3} = 3 + \frac{1}{\alpha_1}, \qquad \text{then} \qquad \frac{1}{\alpha_1} = \frac{-4+\sqrt{37}}{3},$$

and

$$\alpha_1 = \frac{3}{-4+\sqrt{37}} = \frac{4+\sqrt{37}}{7} > 1.$$

On the other hand, $\alpha_1' = \frac{1}{7}(4-\sqrt{37})$ is approximately $-\frac{2}{7}$, so $-1 < \alpha_1' < 0$. Hence α_1 is reduced.

2. From $0 < P < \sqrt{D}$ and $\sqrt{D} - P < Q < \sqrt{D} + P$, see (4.23), it follows that

$$\alpha = \frac{P+\sqrt{D}}{Q} > \frac{Q}{Q} = 1.$$

Since $Q > 0$ and $P - \sqrt{D} < 0$,

$$\alpha' = \frac{P-\sqrt{D}}{Q} < 0.$$

Also, $\sqrt{D} - P < Q$ implies that $(\sqrt{D} - P)/Q < 1$ so that

$$\alpha' = \frac{P-\sqrt{D}}{Q} > -1.$$

3. The totality of expressions of the form $\dfrac{P+\sqrt{43}}{Q}$ where P and Q are integers satisfying condition (4.23) are obtained as follows: If $P = 1$, $\sqrt{43} - 1 < Q < \sqrt{43} + 1$, i.e. $6 \le Q \le 7$; this yields

$$\frac{1+\sqrt{43}}{6}, \quad \frac{1+\sqrt{43}}{7}.$$

If $P = 2$, $5 \le Q \le 8$; this yields

$$\frac{2+\sqrt{43}}{5}, \quad \frac{2+\sqrt{43}}{6}, \quad \frac{2+\sqrt{43}}{7}, \quad \frac{2+\sqrt{43}}{8}.$$

If $P = 3$, $4 \le Q \le 9$; this yields

$$\frac{3+\sqrt{43}}{n}, \qquad n = 4, 5, 6, 7, 8, 9.$$

By the same procedure, for $P = 4, 5, 6$, we find

$$4 + \frac{\sqrt{43}}{k} \text{ with } k = 3, 4, \cdots, 10; \quad \frac{5 + \sqrt{43}}{l} \text{ with } l = 2, 3, \cdots, 11;$$

and $\dfrac{6 + \sqrt{43}}{m}$ with $m = 1, 2, \cdots, 12$.

Set 17, page 110

1. $\alpha = 1 + \sqrt{2} > 1$, $\alpha' = 1 - \sqrt{2} = 1 - 1.414 \cdots$ lies between -1 and 0. Also $1 + \sqrt{2} = 2 + \dfrac{1}{\alpha_1}$, $\alpha_1 = 1 + \sqrt{2} = \alpha$, hence $\alpha = [2, 2, 2, \cdots] = [\overline{2}]$.

2. $\alpha = \sqrt{8} > 1$, $\alpha' = -\sqrt{8}$ does not lie between -1 and 0. $\sqrt{8} = [2, \overline{1, 4}]$.

Set 18, page 112

1. $\dfrac{8 + \sqrt{37}}{9} = 1 + \dfrac{1}{\alpha_1}$,

$$\alpha_1 = \frac{1 + \sqrt{37}}{4} = 1 + \frac{1}{\alpha_2}, \qquad \alpha_2 = \frac{3 + \sqrt{37}}{7} = 1 + \frac{1}{\alpha_3},$$

$$\alpha_3 = \frac{4 + \sqrt{37}}{3} = 3 + \frac{1}{\alpha_4}, \qquad \alpha_4 = \frac{5 + \sqrt{37}}{4} = 2 + \frac{1}{\alpha_5},$$

$\alpha_5 = \alpha_2$, where α_2 is a reduced quadratic irrational. Hence

$$\frac{8 + \sqrt{37}}{9} = [1, 1, \overline{1, 3, 2}] .$$

Notice that α and α_1 are not reduced, but that

$$\alpha_2 = \frac{3 + \sqrt{37}}{7} > 1, \qquad -1 < \alpha_2' = \frac{3 - \sqrt{37}}{7} < 0;$$

hence α_2 is reduced and the continued fraction is periodic from then on.

Set 19, page 118

1. $\sqrt{7} = [2, \overline{1, 1, 1, 4}]$. The convergents are $2/1$, $3/1$, $5/2$, $8/3 = p_4/q_4$ so $p_4 = x_1 = 8$, $q_4 = y_1 = 3$, and

$$x_1^2 - 7y_1^2 = 64 - 7(9) = 64 - 63 = 1.$$

2. $\sqrt{13} = [3, \overline{1, 1, 1, 1, 6}]$. The first five convergents are 3/1, 4/1, 7/2, 11/3, $18/5 = p_5/q_5$, so $p_5 = x_1 = 18$, $q_5 = y_1 = 5$ gives a solution of $x^2 - 13y^2 = -1$. Proceeding to the tenth convergent we find $p_{10}/q_{10} = 649/180$. Thus $x_2 = 649$, $y_2 = 180$ is a solution of $x^2 - 13y^2 = 1$.

Set 20, page 121

1. According to Theorem 4.4, the next two solutions, (x_2, y_2) and (x_3, y_3), are obtained from $x_2 + y_2\sqrt{18} = (x_1 + y_1\sqrt{18})^2$ and

$$x_3 + y_3\sqrt{18} = (x_1 + y_1\sqrt{18})^3.$$

The first relation yields $x_2 + y_2\sqrt{18} = x_1^2 + 18y_1^2 + 2x_1y_1\sqrt{18}$. Since $A + B\sqrt{D} = C + E\sqrt{D}$ if and only if $A = C$ and $B = E$, and since $x_1 = 17$, $y_1 = 4$, we have

$$x_2 = x_1^2 + 18y_1^2 = (17)^2 + 18(4)^2 = 577,$$

$$y_2 = 2x_1y_1 = 2 \cdot 17 \cdot 4 = 136,$$

and

$$x_2^2 - 18y_2^2 = (577)^2 - 18(136)^2 = 1.$$

From the relation for x_3, we have

$$\begin{aligned} x_3 + y_3\sqrt{18} &= x_1^3 + 3x_1^2y_1\sqrt{18} + 3x_1y_1^2 18 + y_1^3 18\sqrt{18} \\ &= x_1^3 + 54x_1y_1^2 + (3x_1^2y_1 + 18y_1^3)\sqrt{18} \end{aligned}$$

so that

$$x_3 = x_1^3 + 54x_1y_1^2, \qquad y_3 = 3x_1^2y_1 + 18y_1^3.$$

If 17 is substituted for x_1 and 4 for y_1, the relation $x_3^2 - 18y_3^2 = 1$ may be verified.

2. According to Theorem 4.5, the next solution of $x^2 - 13y^2 = -1$ is obtained from

$$x_3 + y_3\sqrt{13} = (x_1 + y_1\sqrt{13})^3 = x_1^3 + 39x_1y_1^2 + (3x_1^2y_1 + 13y_1^3)\sqrt{13};$$

$x_1 = 18$, $y_1 = 5$ is the minimal solution which determines the next solution $x_3 = x_1^3 + 39x_1y_1^2$, $y_3 = 3x_1^2y_1 + 13y_1^2$. Solutions (x_2, y_2) and (x_4, y_4) of the equation $x^2 - 13y^2 = 1$ are obtained from

$$x_2 + y_2\sqrt{13} = (x_1 + y_1\sqrt{13})^2 \text{ and } x_4 + y_4\sqrt{13} = (x_1 + y_1\sqrt{13})^4.$$

The computations are left to the diligent reader.

3. Table 2 indicates that $u_1 = 1$, $v_1 = 1$ is the minimal solution of $u^2 - 2v^2 = -1$. This yields $u_1 = 1$, $v_1 = n_1 = 1$, $m_1 = 2$, $x_1 = 4$,

$y_1 = 3,\ z_1 = 5;$

$$x_1^2 + y_1^2 = 3^2 + 4^2 = 25 = z_1^2.$$

Other solutions of $u^2 + 2v^2 = \pm 1$ are obtained from

$$u_k + v_k \sqrt{2} = (u_1 + v_1 \sqrt{2})^k = (1 + \sqrt{2})^k \qquad \text{for } k = 2, 3, \cdots.$$

Thus, for $k = 2$, $u_2 + v_2 \sqrt{2} = 3 + 2\sqrt{2}$, and $u_2 = 3$, $v_2 = n_2 = 2$, $m_2 = 5$, $x_2 = 20$, $y_2 = 21$, $z_2 = 29$;

$$x_2^2 + y_2^2 = 841 = z_2^2.$$

For $k = 3$, $u_3 + v_3 \sqrt{2} = (1 + \sqrt{2})^3 = 7 + 5\sqrt{2}$, and $u_3 = 7$, $v_3 = n_3 = 5$, $m_3 = 12$, $x_3 = 120$, $y_3 = 119$, $z_3 = 169$;

$$x_3^2 + y_3^2 = 14400 + 14161 = 28561 = z_3^2.$$

For $k = 4$, $u_4 + v_4 \sqrt{2} = (1 + \sqrt{2})^4 = 17 + 12\sqrt{2}$, and $u_4 = 17$, $v_4 = n_4 = 12$, $m_4 = 29$, $x_4 = 696$, $y_4 = 697$, $z_4 = 985$;

$$x_4^2 + y_4^2 = 484416 + 485809 = 970225 = z_4^2.$$

4. As explained in the statement of Problem 3, Set 20, page 121, the lengths of the sides may be written

$$x = m^2 - n^2, \qquad y = 2mn, \qquad z = m^2 + n^2$$

where m and n are positive integers, $m > n$. Therefore

$$\tan\frac{\theta}{2} = \frac{\sin\theta}{1 + \cos\theta} = \frac{y/z}{1 + (x/z)} = \frac{2mn/(m^2 + n^2)}{1 + (m^2 - n^2)/(m^2 + n^2)} = \frac{n}{m}.$$

If we could find sequences of integers $n_1, n_2, \cdots$ and $m_1, m_2, \cdots$ such that the ratios n_1/m_1, $n_2/m_2, \cdots$ approach $1/\sqrt{3}$, then $\theta/2$ would approach 30° and θ would approach 60°. To find these sequences, we convert $\sqrt{3}$ into the continued fraction

$$\sqrt{3} = 1 + \frac{1}{1+}\,\frac{1}{2+}\,\frac{1}{1+}\,\frac{1}{2+}\cdots$$

and let m_i and n_i be the numerator and denominator, respectively, of the convergent c_i. We find

$$m_i: \quad 2, 5, 7, 19, 26, 71, \cdots$$

$$n_i: \quad 1, 3, 4, 11, 15, 41, \cdots$$

and corresponding triangles with sides (3, 4, 5), (16, 30, 34), $\cdots$. The sixth triangle has sides (3360, 5822, 6722) and its angle θ is between 60° and 61°, but much closer to 60°.

Set 21, page 130

1. The first six convergents to $\alpha = \frac{1}{3}(1 + \sqrt{10}) = 1.3874 \cdots$ are $\frac{1}{1}$, $\frac{3}{2}$, $\frac{4}{3}$, $\frac{7}{5}$, $\frac{18}{13}$, $\frac{25}{18}$, and the convergent $\frac{7}{5}$ satisfies the inequalities (5.6); note that $\frac{1}{1}$ also satisfies (5.6).

2. F_5: $$\frac{0}{1},\ \frac{1}{5},\ \frac{1}{4},\ \frac{1}{3},\ \frac{2}{5},\ \frac{1}{2},\ \frac{3}{5},\ \frac{2}{3},\ \frac{3}{4},\ \frac{4}{5},\ \frac{1}{1}.$$

3. In F_2, $\alpha = .387 \cdots$ lies between $\frac{0}{1}$ and $\frac{1}{2}$. Of the numbers
$$\frac{0}{1}, \qquad \frac{0+1}{1+2} = \frac{1}{3}, \qquad \text{and} \qquad \frac{1}{2},$$
the first satisfies (5.6). The other two come close but do not make it.

4. $x \sim y$ because $(-10)(-5) - (7)(7) = 1$.
$$x = [\overline{1}] \quad \text{and} \quad y = [-2, 1, 1, 4, \overline{1}] = \frac{-169 - \sqrt{5}}{118}.$$

5. (i) Since $x = \dfrac{ax + b}{cx + d}$, with $a = 1,\ b = 0,\ c = 0,\ d = 1$, we have $ad - bc = 1 - 0 = 1$. Hence $x \sim x$.

(ii) If $x = \dfrac{ay + b}{cy + d}$ and $ad - bc = \pm 1$, then
$$y = \frac{-dx + b}{cx - a} = \frac{Ax + B}{Cx + D},$$
where $AD - BC = ad - bc = \pm 1$. Hence $y \sim x$.

(iii) Since $x \sim y$ and $y \sim z$, we can write
$$x = \frac{ay + b}{cy + d} \qquad \text{and} \qquad y = \frac{a'z + b'}{c'z + d'},$$
where, respectively, $ad - bc = \pm 1$ and $a'd' - b'c' = \pm 1$. Then
$$x = \frac{a\left(\dfrac{a'z + b'}{c'z + d'}\right) + b}{c\left(\dfrac{a'z + b'}{c'z + d'}\right) + d} = \frac{(aa' + bc')z + ab' + bd'}{(ca' + dc')z + cb' + dd'} = \frac{Az + B}{Cz + D},$$

and

$$\begin{aligned} AD - BC &= (aa' + bc')(cb' + dd') - (ab' + bd')(ca' + dc') \\ &= aa'dd' + bb'cc' - a'bcd' - ab'c'd \\ &= a'd'(ad - bc) - b'c'(ad - bc) \\ &= (ad - bc)(a'd' - b'c') \\ &= \pm 1. \end{aligned}$$

Set 22, page 133

1. $\sqrt{29} = [5, \widetilde{2, 1, 1, 2, 10}]$, so we must calculate α_3: $\alpha_3 = \dfrac{2 + \sqrt{29}}{5}$, hence $P = 2$, $Q = 5$, and $29 = 2^2 + 5^2$.

2. $\sqrt{433} = [20, \widetilde{1, 4, 4, 2, 2, 1, 3, 13, 1, 1, 1, 1, 13, 3, 1, 2, 2, 4, 4, 1, 40}]$, so we must calculate α_{11}:

$$\alpha_{11} = \frac{12 + \sqrt{433}}{17}, \text{ hence } P = 12,\ Q = 17 \text{ and } 433 = 12^2 + 17^2.$$

3. Since $\sqrt{2}$ is irrational it is impossible to find two integers a and b such that

$$\sqrt{2} = \frac{a}{b}, \qquad \text{or such that} \qquad a^2 = 2b^2 = b^2 + b^2.$$

On the other hand

$$\sqrt{2} = 1 + \frac{1}{2+}\,\frac{1}{2+}\,\frac{1}{2+}\cdots$$

and the convergents to this continued fraction are

$$\frac{1}{1}, \frac{3}{2}, \frac{7}{5}, \frac{17}{12}, \frac{41}{29}, \frac{99}{70}, \ldots, \frac{p}{q}, \ldots;$$

we always have

$$p^2 - 2q^2 = \pm 1 \qquad \text{or} \qquad p^2 \pm 1 = 2q^2 = q^2 + q^2.$$

Hence the second part of the problem can be solved by values of p and q such that $p^2 + 1 = 2q^2$, or $p^2 - 1 = 2q^2$.

References

The books listed below either contain chapters on continued fractions, or deal with subject matter that has been referred to in the text. No attempt has been made to compile a complete bibliography. The standard treatise on continued fractions is Perron's *Kettenbrüche*, but this book is for the specialist. The only extended account of the subject in English is that given in Vol. II of Chrystal's *Algebra*, an old-fashioned yet still valuable text. The book by Davenport is very good reading, for he gets to the heart of the matter quickly and with very little fussing.

1. Edwin Beckenbach and Richard Bellman, *An Introduction to Inequalities*, New Mathematical Library 3, New York: Random House, Inc., 1961.

2. G. Chrystal, *Algebra*, vol. II, Edinburgh: Adam and Black, 1889; reprinted, New York: Chelsea, 1959.

3. H. Davenport, *The Higher Arithmetic*, London: Hutchinson's University Library, 1952.

4. L. E. Dickson, *History of the Theory of Numbers*, vols. I, II, III, Washington: Carnegie Institute of Washington, 1919; reprinted, New York: Chelsea, 1950.

5. G. H. Hardy and E. M. Wright, *An Introduction to the Theory of Numbers*, 4th ed., Oxford: Clarendon Press, 1960.

6. Sir Thomas L. Heath, *Diophantus of Alexandria, A Study in the History of Greek Algebra*, 2nd ed., Cambridge: Cambridge University Press, 1910.

7. W. J. LeVeque, *Topics in Number Theory*, vols. I and II, Reading, Mass.: Addison-Wesley, 1956.

8. Ivan Niven, *Irrational Numbers*, Carus Monograph 11, New York: John Wiley and Sons, 1956.

9. Ivan Niven, *Numbers: Rational and Irrational*, New Mathematical Library 1, New York: Random House Inc., 1961.

10. O. Ore, *Number Theory and Its History*, New York: McGraw-Hill, 1949.

11. Oskar Perron, *Die Lehre von den Kettenbrüchen*, Leipzig and Berlin: Teubner, 1929.

12. Raphael M. Robinson, *Unsymmetrical Approximations of Irrational Numbers*, Bulletin American Math. Soc., vol. 53 (1947) pp. 351–361.

13. D. E. Smith, *History of Mathematics*, vols. I and II, New York: Dover Publications, Inc., Reprint, 1958.

14. H. S. Wall, *Analytic Theory of Continued Fractions*, New York: D. Van Nostrand Company, Inc., 1948.

15. Leo Zippin, *Uses of Infinity*, New Mathematical Library 7, New York: Random House, Inc., 1962.

Index